Palgrave Studies in Educational Futures

Series Editor
jan jagodzinski
Department of Secondary Education
University of Alberta
Edmonton, Canada

The series Educational Futures would be a call on all aspects of education, not only specific subject specialists, but policy makers, religious education leaders, curriculum theorists, and those involved in shaping the educational imagination through its foundations and both psychoanalytical and psychological investments with youth to address this extraordinary precarity and anxiety that is continually rising as things do not get better but worsen. A global de-territorialization is taking place, and new voices and visions need to be seen and heard. The series would address the following questions and concerns. The three key signifiers of the book series title address this state of risk and emergency:

1. **The Anthropocene**: The 'human world,' the world-for-us is drifting toward a global situation where human extinction is not out of the question due to economic industrialization and overdevelopment, as well as the exponential growth of global population. How to we address this ecologically and educationally to still make a difference?
2. **Ecology**: What might be ways of re-thinking our relationships with the non-human forms of existence and in-human forms of artificial intelligence that have emerged? Are there possibilities to rework the ecological imagination educationally from its over-romanticized view of Nature, as many have argued: Nature and culture are no longer tenable separate signifiers. Can teachers and professors address the ideas that surround differentiated subjectivity where agency is no long attributed to the 'human' alone?
3. **Aesthetic Imaginaries**: What are the creative responses that can fabulate aesthetic imaginaries that are viable in specific contexts where the emergent ideas, which are able to gather heterogeneous elements together to present projects that address the two former descriptors: the Anthropocene and the changing modulating ecologies. Can educators draw on these aesthetic imaginaries to offer exploratory hope for what is a changing globe that is in constant crisis?

The series Educational Futures: Anthropocene, Ecology, and Aesthetic Imaginaries attempts to secure manuscripts that are aware of the precarity that reverberates throughout all life, and attempts to explore and experiment to develop an educational imagination which, at the very least, makes conscious what is a dire situation.

Miranda Matthews
Editor

Ecologies in Practice and Learning

Arts Interventions in the Earth Crisis

palgrave
macmillan

Editor
Miranda Matthews
Educational Studies
Goldsmiths, University of London
London, UK

Palgrave Studies in Educational Futures
ISBN 978-3-031-82337-4 ISBN 978-3-031-82338-1 (eBook)
https://doi.org/10.1007/978-3-031-82338-1

This Palgrave Macmillan imprint is published by the registered company Springer Nature Switzerland AG.
The registered company address is: Gewerbestrasse 11, 6330 Cham, Switzerland

The original version of the book has been revised. A correction to this book can be found at https://doi.org/10.1007/978-3-031-82338-1_21

Acknowledgements

The editor Miranda Matthews would like to acknowledge all of the authors and projects that are included in this book, their geopolitical contexts, the participating communities, other-than-human lifeforms, and more-than-human lifeforces that are all important to ecologies in practice and learning.

Praise for *Ecologies in Practice and Learning*

"The reference to an 'Earth Crisis' in this book's title reminds its readers that we are compelled to think globally. Even the smallest details have become representative of much larger predictive models and predicaments, whether we take responsibility for them or not. In this overwhelming process of futures thinking, the scope of art and education is extended to reconceptualise everything we observe, touch, eat and, more likely than not, take for granted. This is what makes this book such a vital and timely exploration of reparative practices and theoretical insights that not only challenge traditional educational and artistic paradigms but advocate for a new understanding of human and other-than-human interconnection. Its range of international, artistic, collective and further insightful perspectives make this book essential reading for anyone committed to creatively reshaping our possible futures."

—Raphael Vella, *University of Malta*

Contents

Notes on Contributors

Andrew Amondson is an American-born artist and filmmaker living in Berlin. His work in film includes collaborations with Wim Wenders, the Berlinale, and as a creative strategist for Netflix. He co-founded Playful Nature, creating play-based art installations in Berlin. He has been a member of Studio Olafur Eliasson since 2017 and a lead artist with Forest of Imagination since 2020. He creates living installations of care, nature-connectedness, and the ecologies of imagination. His Living Tree Forest installation was the recipient of the Inspire Future Generations Award in London, 2022. Andrew is a fellow of the Royal Society for Arts.

Dennis Atkinson is Professor Emeritus at Goldsmiths University of London, and a visiting professor at the Universities of Porto, Gothenburg and Barcelona. He was Principal Editor of *The International Journal of Art and Design Education* (2002–2009). He has published regularly in international journals, contributed to edited collections and published several books, including *Art, Equality and Learning: Pedagogies Against the State*, 2013), *Art, Disobedience and Ethics: Adventures of Pedagogy* (2017), and *Pedagogies of Taking care: Art, Pedagogy and the Gift of otherness* (2022). He received The Ziegfeld Award (2015) from The United States Society for Education through Art, for outstanding international contributions to art in education. (E-mail: d.atkinson@gold.ac.uk).

Landing Bassene was brought up in the South of Senegal and attended Cheikh Anta DIOP University in Dakar and graduated in English literature and Documentary Information Sciences. He worked for the Korea International Cooperation Agency, then trained as a secondary school teacher. With a grant from the Italian Cooperation Agency, he worked on an international course in Applied Anthropology for Development Processes at La Sapienza University in Rome. Back in Senegal, Landing worked at the newspaper *Le Quotidien*, before joining the Masters in Learning and Communication in Multilingual and Multicultural Contexts at the University of Luxembourg.

Gabriele Budach is Associate Professor for Teacher Education, Learning and Diversity at the University of Luxembourg. She is a sociolinguist and educational researcher, interested in multilingualism, communication and learning. Her early research and teaching have been focusing on multilingual literacies, adult literacy in francophone Canada and German-Italian biliteracy in primary education in Germany. Working with Inuit women in Ottawa/Canada on culturally appropriate pedagogies, she was introduced to indigenous epistemologies and collaborative ethnography. In her recent work, she uses multimodal, arts-based methods to explore issues of racism, environmental education and justice, based in frameworks of decoloniality and new materialism.

Kevin Campbell Davidson is a lecturer and researcher in the Department of Educational Studies at Goldsmiths College, University of London. Kevin is interested in the role of play, creativity, and artistic movement practices in learning. As an associate of Climate Museum UK, Kevin conducts research into the role of the arts to support young people with eco-anxiety and is consulted by organisations who are producing learning resources around planetary health. Kevin has recently been awarded a Churchill Fellowship to explore how universities are adapting their learning environments in response to the climate and ecological crisis.

Sudip Chakroborthy teaches theatre and performance at the University of Dhaka, Bangladesh. He trained in South Asian traditional performance culture, both as a performer and as a director. He has been actively involved with numerous theatre creations, workshops, seminars, conferences, and festivals. His theatre direction, designs, and lectures have been performed on stages in Bangladesh, India, Bhutan, South Korea, Iran, France, England, Scotland, Wales, the United States, and Canada. Sudip is a regular speaker at international conferences. He has written in journals in Bangladesh and the United Kingdom. He has co-authored a book entitled *Traditional Games of Bangladesh* published by UNESCO.

Francis Gilbert is Senior Lecturer in Educational Studies at Goldsmiths, UK, where he is the head of Subject and head of the Masters' programmes, head of the MA Creative Writing and Education and academic co-director of the Connected Curriculum. He has been the Principal Investigator of the Parklife Project since 2021, which has received a Goldsmiths' Strategic Fund grant (2021–2023) and British Academy Social Sciences, Humanities and the Arts for People and the Economy (SHAPE) funding (2022–2024). www.francisgilbert.co.uk

Emese Hall is Senior Lecturer in Art Education and the former Director of the PGCE Primary Programme at the University of Exeter, UK. Emese has worked in teacher education since 2005. She researches the value of professional learning communities, artist-teacher practice, and the potential of art education to address the Earth Crisis. Among other honours, Emese was Vice President of the UK National Society for Education in Art and Design (NSEAD) and Subject Lead for Primary Art, Craft and Design at Oak National Academy. Emese co-edited the Pedagogy Volume of *The International Encyclopaedia of Art and Design Education* (Wiley-Blackwell, 2019).

Penny Hay is Professor of Imagination and Reader in Creative Teaching and Learning at Bath Spa University. Hay is an artist, educator and researcher, and Founding Director of House of Imagination. Her doctoral research focused on children's learning identity as artists, and she has worked internationally on arts education research projects. She is the strand leader for Creative Pedagogy in the Policy, Pedagogy and Practice Research Centre, and co-chairs the eARTh research group. Hay was awarded an Honorary Fellowship at Arts University Plymouth and a Fellowship in Imagination at the Centre for Future Thinking; she is a National Teaching Fellow and Fellow of the Chartered College of Teaching. Twitter/X handles @PenAHay @forestimaginatn @houseimagninatn @BathSpaUniLinkedin https://www.linkedin.com/in/dr-penny-hay-5982b01a/

Aminul Hoque lectures in the Educational Studies Department at Goldsmiths. He gained his doctorate from Goldsmiths in 2011, and his in-depth ethnographic research forms the basis of his book *British-Islamic Identity: Third-Generation Bangladeshis from East London* (2015). Aminul's research focuses on issues of multicultural Britain, identity, social justice, youth policy, religion, culturally responsive pedagogies, inclusive education, race relations, and Islamic feminism. Aminul is also a freelance journalist and broadcaster. His radio documentary *Islamic Pride* was shortlisted for the prestigious Sony Awards (2004). In 2020, he presented the television documentary *A Very British History: British Bangladeshis* on BBC4.

Graham Jeffery is the director of the Protracted Crisis Research Centre at the University of the West of Scotland, UK. He graduated in Music from the University of Cambridge in 1991, and has worked in secondary, further, and higher arts education. Graham spans participatory and community arts practices, creative pedagogies, cultural policy, and urban and community development. He worked as a teacher, composer, and facilitator, with London International Festival of Theatre, BBC Scottish Symphony Orchestra and East London Dance; also in policy advice and consultancy with Creative Scotland and the Department for Culture, Media and Sport. Graham leads action research projects with diverse global communities.

Erin Manning is Professor of Fine Arts and Philosophy at Concordia University. Recent books include For a Pragmatics of the Useless (2020), Out of the Clear (2022) and The Being of Relation (2025). She works at the intersection of the three ecologies—the environmental, the conceptual and the social—with an emphasis on the aesthetico-political (3ecologies.org). Her artistic practices explore this transversality—a related exhibition is entitled *100 Acres* (Richard Saltoun Gallery, London 2024).

Miranda Matthews has contributed to practice, teaching, and research in arts education and related fields for more than 20 years. Initially Matthews taught in schools and further education (2004–2014), also completing her PhD in Educational Studies at that time (Goldsmiths, 2012). Matthews joined the Department of Educational Studies at Goldsmiths, University of London, UK, in 2016. She became head of the Centre for Arts and Learning (CAL) in 2019. Matthews curates programmes of public facing research projects and events. The assemblage of annual themes has enabled focused, responsive interdisciplinary research development for arts educators. Matthews has created and encouraged *Ecologies in Practice* research since 2021, in a range of national and international Centre for Arts and Learning events and projects.

Contributors to the 2021–2023 research activities were invited to write a chapter for *Ecologies in Practice and Learning: Arts Interventions in the Earth Crisis.* Matthews edited this book to assemble the voices of significant international contributors, to gather diverse inspirational practices, and to represent relevant and innovative arts pedagogies. To further *Ecologies in Practice* research, Matthews was Principal Investigator (PI) for an Arts and Humanities Research Council (AHRC) funded research project (2022–2023). The cultural partner for this project was Climate Museum UK.

Matthews continues to collaborate on practice research that explores and validates ecologies in the arts, and interdisciplinary practice as forms of research inquiry. Matthews currently teaches on arts in education undergraduate and postgraduate degree programmes. She also supervises UK and international PhD students in collaboration with cultural organisations and international higher education institutions. Matthews' gathering range of publications include recent writing that connects posthumanist, affective, and imaginative ways of realising more sustainable learning environments, with newly generated developments in theory and practice. Matthews' book, *Arts Methods for the Self-Representation of Undergraduate Students* (2023), presents her research on how transitions into university cultures can be assisted by arts practice. She is also an editor of the *International Journal of Art and Design Education* (iJADE, 2019–present). In addition to articles in academic journals—such as 'Arts Practice as the Daily Extraordinary' (iJADE 2024)—Matthews has contributed chapters to collections such as *The Bloomsbury Handbook of Contemporary Philosophy of Education* (Eds. J. Baldacchino and H. Saeverot 2024) and *Professional Learning for Artist Teachers: Pedagogy, Practice and Partnership in UK Contexts* (Ed. R. Payne, 2024). Goldsmiths Staff Webpage: https://www.gold.ac.uk/educational-studies/staff/matthews-miranda/ Centre for Arts and Learning Webpage: https://www.gold.ac.uk/cal/

Ben McDonnell is an artist researcher, his background as a musician informs his visual arts practice that materially questions how we work together—raising questions around stability, support and structure. His BA was in music composition and performance at the Leeds College of Music, and his MA was in Photography from the Royal College of Art. Ben's current doctoral research at Goldsmiths, University of London, forms relationships between sound image and pedagogy, focusing on graphic notation to score and transcribe learning events. He is also Senior Lecturer in Fine Art at Norwich School of the Arts.

Bridget McKenzie is a researcher, trainer and artist working to engage people with the Earth Crisis. She presents and publishes internationally on Regenerative Culture. Her career includes being Tate's Education manager and the British Library's Head of Learning. In 2006, she founded Flow Associates, a research consultancy in cultural learning. In 2019, she founded Climate Museum UK, a creative collective helping people make sense of environmental issues. She also co-founded Culture Declares, an international movement of cultural workers declaring a climate and ecological emergency. She experiments with creative activism in her local city of Norwich, in Norfolk. https://bridgetmckenzie.uk/

Terri Newman is a collaborative artist, teacher and practice-based researcher in Arts and Learning studies. Their collaborative practice investigates the pedagogy of collaboration, carving space for collaborative practices within the art and design curriculum. Terri has developed collaborative methodologies for the post-sixteen classroom as a classroom teacher, faculty leader and senior leader, working towards bridging the gap between post-sixteen and higher educational study for young people from the most disadvantaged backgrounds.

Molly Pardoe is an artist, researcher and a PhD student at the Centre for Arts and Learning, Goldsmiths University of London. Her practice-based painting research sits at the intersection between pedagogies of art and transnational studies. She majored in psychology and communication at the University of Mary Washington and received her MSc in psychosocial studies in education at Birkbeck College. She has also facilitated pedagogical research at Loris Malaguzzi International Centre. She currently lectures and organizes research groups in practice-based research methodologies at Goldsmiths and co-organises a multi-institutional collaborative research group that interrogates the pedagogies of collective practices.

Ben Parry, an artist, curator, and practice-led researcher based in Bristol, is Reader in Art, Environment, and Social Practice at Bath Spa University, UK. He received his BA in Environmental Art from Glasgow School of Art, MA in Urban Planning from the University of Liverpool, and PhD from the University of the West of Scotland. Working internationally, his projects employ a wide range of mediums including film, photography, light, sound, and machines. His research and practice explore the intersections of art intervention, urbanism and activism, and the self-organising systems of the user-generated city.

Henrietta Patience is a lecturer in Art and Design Education teaching on the PGCE Secondary and Primary programmes at Goldsmiths University of London. Henrietta has taught Art and Design and Film Studies in South East London Secondary schools for over 20 years and continues to support curriculum development in schools, with research interests in inclusion and aspiration towards creative pathways and sustainability.

Martina Reuter, WochenKlausur the Vienna based artists' collective WochenKlausur has worked since 1993, on invitation from international art institutions, to improve sociopolitical deficiencies. The collective develops small time and site specific proposals aimed at small, but nevertheless effective changes that are taken into action. www.wochenklausur.at

Stephania Rovira Ochoa graduated from the Universidad Pedagógica Nacional of Bogotá-Colombia, with a BA in Education and Foreign Language Teaching. She worked as a teacher from 2016 to 2019 before moving to Euskirchen, Germany, where she worked at a school for children with special needs. In 2020, she relocated to Luxembourg to pursue a Masters in Learning and Communication in Multilingual and Multicultural Contexts. She is currently a research and development specialist at the University of Luxembourg, working on a social project around the topic of intercultural third places.

Sharmila Samant is an Indian visual artist and arts educator. Sharmila played a key role in developing the Compound 13 Lab project. She graduated in Sculpture from Sir J. J. School of Arts, followed by a Diploma in Interior Design, a Fellowship at the Kanoria Center for Arts, Ahmedabad, and artist-residencies at Rijksakademie, Amsterdam, and Gasworks, London. Samant co-founded the Mumbai-based collective Open Circle that engaged with contemporary socio-political issues via integrated theory and practice. Negotiating in private and public spheres, Samant has engaged in collaborative, participatory art projects with various communities while exploring ideas of exchange, accessibility, and authorship.

Trish Scott is a Lecturer in Arts and Cultural Management Education in the Department for Culture, Media and Creative Industries at King's College London. She has extensive experience in the UK arts sector and has previously held curatorial positions at Goldsmiths Centre for Contemporary Art, Tate, and Turner Contemporary. Scott's practice exists at the intersection of experimental education, curating, and research, with a focus on how participative approaches can challenge and dismantle traditional organisational hierarchies and systems of value and create space for multiple voices.

Gohar Sharoyan graduated in Linguistics, Translation and Intercultural Communication at Yerevan Brusov State University in Armenia. Gohar holds a Masters in Learning and Communication in Multilingual and Multicultural Contexts from the University of Luxembourg. Having worked in radio production at a local community station, Gohar now shares the art of audio storytelling with students in a course on 'Podcasting' at the University of Luxembourg. In addition to teaching, she has been working as a Research and Development Specialist on several projects facilitating outreach activities, including stop motion animation workshops for storytelling and self-expression.

Jenn Steverson is a textile artist and independent scholar based in Central Texas. Her practice explores the complexity of Black geographies through quilt making, botanical dyeing, archival photographs, and texts. She frames Black craft traditions as placemaking practices that have withstood displacement. In 2021 Jenn was a Virtual Artist+ Scholar Resident with the Black Botany Studio at UC Santa Cruz. She was a Creator-Maker Fellow at the Winterthur Museum and Library in 2022. Her work was recently published in a special edition of the Arrow Journal, Undisciplined Archives: Dreaming Across Black Geographies, 2023.

Anna Stewart holds a Master's Degree in Psychology Research from Goldsmiths University, UK, and is currently a psychology researcher at i2 Media Research Ltd. She specialises in environmental psychology and the psychology of wellbeing, with her work spanning the technology, sustainability and creative arts sectors. From 2022 to 2023, Anna was a creative co-researcher on the Parklife Project, where she taught and collaborated on 360-degree filming and editing, integrating these skills with traditional psychology research techniques.

Chris Turner OBE is an honorary lecturer at the School of Education, University of Exeter, and an associate fellow at the Centre for Climate Change and Sustainability Education, University College London. His research and writing interests centre on the entanglement between aesthetics and ecology from which he has developed the theoretical concepts of aesthoecology. As an eco-philosopher he engages with issues of relationality and anthropogenic environmental degradation. He is a member of the Creativity and Emergent Educational Futures Network at the University of Exeter, a Fellow of the Royal Society of Arts and a Fellow of the Linnean Society.

Roberto Andrés Zamora Figueroa and Juan Pablo Zamora Figueroa are filmmakers based in Berlin who have taken ecological filmmaking to the extent of every possibility they can realise. The Zamora brothers argue for film production and distribution processes to become sustainably aware of the Earth Crisis in all decisions made, from wardrobe to location and lighting, so that every element becomes waste-free, mitigating carbon emissions and caring of the Earth's ecosystems. The Zamoras invite us to accompany them in visualising a day on a sustainable film set. They also share their guide for Regenerative Filmmaking as a transferable code of practice for future international ecological films.

Abbreviations

ACORN	Association of Community Organisations for Reform Now
AE	Art Education
AQA	The Assessment and Qualifications Alliance
CAL	Centre for Arts and Learning
CARE	Connecting Art with Real Life Issues
CARICOM	Caribbean Community
CCA	Goldsmiths Centre for Contemporary Art
C13	Compound 13
ESFA	The European Food Safety Authority
HEI	Higher Education Institution
CED	Climate Emergency Declaration
CMUK	Climate Museum UK
CO_2	Carbon Dioxide
COP26	26th Conference of the Parties
GG	Global Goals
GRU	Germplam Resource Unit
IEA	International Energy Agency
InSEA	International Society for Education Through Art
IPCC	Intergovernmental Panel on Climate Change
IUCN	International Union for Conservation of Nature
LGBTQ+	Lesbian, Gay, Bisexual, Trans, Queer Plus
MFA	Masters in Fine Art
NESTA	National Endowment for Science, Technology and the Arts
NGO	Non-Governmental Organisation
ODA	Official Development Assistance
PLC	Professional Learning Community
RT	Reciprocal Teaching

SE	Sustainability Education
SEN	Special Educational Needs
SOS-UK	Students Organising for Sustainability UK
STEAM	Science, Technology, Engineering, Art and Mathematics
STG	Systems Thinking Games
Ofsted	Office for Standards in Education
UNESCO	United Nations Educational, Scientific and Cultural Organisation
UNICEF	The United Nations Children's Fund
YCR	Young Co-Researchers

List of Figures

CHAPTER 1

Introduction: Ecologies in Practice and Learning: Arts Interventions in the Earth Crisis

Miranda Matthews

Introduction

In these times of drastic change for the Earth's biosphere, arts education practice research is amplifying multiple voices for bringing the Earth to a better place for as many billions of years as possible, in as short a time as possible. Our information suggests that there is little time to prevent the planet being robbed of all of its resources, before this hollowing extractivism takes us beyond any temperate climate or nature, to that extremely sad void of no return. This hopelessness is what the collected authors of this book passionately want to fight against, in the here and now—for a more flexible sense of what the future could be. Routes via arts practice, education and research are collected here in a constellation of international

M. Matthews (✉)
Educational Studies, Goldsmiths, University of London, London, UK
e-mail: M.Matthews@gold.ac.uk

M. Matthews (ed.), *Ecologies in Practice and Learning: Arts Interventions in the Earth Crisis*, Palgrave Studies in Educational Futures, https://doi.org/10.1007/978-3-031-82338-1_1

approaches that are making this really urgent work more engaging, accessible, enjoyable, entertaining and even playful.

The approaches presented here seek to be resourceful in energising, tangible, adventurous, imaginative, aesthetically interesting forms of learning that support all participants to speak out and materialise climate wellbeing and justice. There are numerous arts activities and professionalised forms of ecological interaction brought together for ecologies in practice. This book celebrates the equality of interventions made in non-formal and formal places of learning, in universities, schools, communities, galleries and museums.

Ecological ontologies as ways of being, or 'ecosophies', place a higher social value on the intra-actions that reach beyond industrialised 'Enlightenment' senses of what it means to learn, to grow and to progress, into arts practice as a site for exchange, connection and collaboration. Posthumanist connections emerge in all areas of ecologies in practice, as authors seek to problem-pose the human, towards intra-subjectivities and rhizomatic lines of flight.

Imaginative leaps involve some capacity for risk and review in practice and research. Arts practice that enables risk taking within supportive spaces, to heal the anxiety that can immobilise productive ecological action, is here presented as it can bring about cohesion in diversity. The diversity that we seek to champion for humanity is situated in awareness of the urgent need to respect and cherish the biodiversity that the Earth is currently losing.

Sometimes other-than-humans may be seen to visibly accompany us on this step by step trek to intra-subjectivity, on our walks, even at our times of creative solitude, but it is the togetherness of difference, and the awakening to the vast array of biodiverse existence that is still here with us, that education brings into focus. Also presented here are ways that arts practice develops care for those who have been excluded from education, and care for imaginative capacities that have historically been wasted. Eco-arts pedagogies are reaching for an expanded view of humanity, including humans and other than humans in all walks of life.

This book brings together voices that offer amplified harmonics of ecological connectivity and critical tools to be passed onwards, for equipping all generations to bring balance back to the Earth. Readers are invited to reflect and navigate to and from formal and non-formal learning settings, inviting more discursive exchange among these different axes of activity. Chapters include the projects of artist educator researchers who gather an extensive range of participants, and also the small group dynamics of collaborative studio ecologies.

The ecological interventions in practice and learning begin with research that addresses climate injustice, sharing the vitality of arts projects that create equality and cultural exchange with the Global South. Then the book presents ecological practice research developed at UK and European universities with schools and local communities, to enable agency, voice, protest and self-expression for young people. Zoning in on specific arts pedagogies, the book then observes collaborative ecologies in practice in microcosms of activity that have macroscopic implications. The next section of *Ecologies in Practice and Learning* introduces inclusive 'possitopian' and imaginative rewilding among museums and galleries. Then our inquiry into how educational arts practice can assist mindful and embodied Earth awareness moves to important works, made by global artists and arts collectives, that are challenging wasteful and destructive cultural practices, and alerting the artworlds to the need for conscientious reparative action. Finally our collection concludes with new ecosophies, materialities and aesthetic imaginaries from contributors who are well known for weaving theory and practice.

As the editor of this collection, I would like to thank all the authors and their collaborators, participants and audiences for sharing projects that infuse positive, hopeful, protesting and celebratory learning in the arts, affirming the multiplicity of eco-pedagogies. We are with the Earth and all of its incredible generative dimensions. We are also with young people and future generations, taking on struggles and facing dilemmas that will continue to be voiced in the volume of arts practice happening in the galactic urgency of the situation.

PART I

Addressing Climate Injustice

Climate injustice needs to be of increasing concern to educators in the wealthy industrialised world. In this section, the book connects with projects that seek to address climate injustice with an expanded view of what arts practice can become. These chapters are co-written between collaborators working in the Global South and researchers based in Europe. Authors present research interventions that are developed over years and are ongoing, in particular localities in Africa and Asia. Here there are celebrations of decolonial ecologies, and practice research with people who are particularly affected by global warming.

Landing Bassene, Stephania Rovira Ochoa, Gohar Sharoyan and Gabriele Budach share the journey of their research into how embodied learning through stop motion animation can make decolonial eco-art interventions in the ecological crisis of Eloubalire, Senegal. Bassene has experienced the destruction of his community's ecological lifelines, by the effects of over-industrialisation in the Global North. In the processes of their research, a bond is formed in which Rovira Ochoa, Sharoyan and Budach—practitioners based in Luxembourg—no longer feel distanced from Bassene's cultural experiences. Yet although they have experienced playful arts practice together, they are still staggered by the scale of the climate upheaval they encounter.

Graham Jeffery, Ben Parry and Sharmila Samant present their collaboration between projects Compound 13 Lab and ACORN in the mega-city of Mumbai, to engage with India's 'circular economy' of waste disposal and recycling. The climate aware in the Global North call out for more recycling, but as yet not many are contributing to the lives of the people who do the actual work of recycling in the Global South. Jeffery, Parry and Samant have made long term arts interventions to improve the life quality of waste workers and their families, to increase educational opportunities, and to share findings about how artists can create ethical 'sustainable urbanism'.

Aminul Hoque and Sudip Chakroborthy are researchers who were born in Bangladesh, one of the countries identified for Official Development Assistance (ODA) by the United Nations—as is Senegal. Bangladesh is extremely affected by climate change. This chapter takes the form of a conversation between Hoque and Chakroborthy, about the ways in which the performing arts can be empowering in the Earth Crisis. Chakroborthy directed a production, *Dokkhina Sundari*, in 2014 to raise awareness of the effects of deforestation in Bangladesh, including the devastation of the Bengal tigers. Hoque and Chakroborthy discuss outcomes of this project, and the context for further performing arts interventions that Chakroborthy continues to make.

Countries in the Global South whose indigenous ecosystems have been severely affected by the impacts of climate change, and the production of colonial cash crops, are also sent the waste of the West. Livelihoods that are already suffering in the South deserve more urgent attention in all areas of education. Arts interventions made in the projects represented here are intently working to improve quality of life, and to develop deep bonds that move beyond superficial reflections on the multi-layered issues that the West needs to make reparations for.

CHAPTER 2

Saving Eloubalire: An Educational Journey of Environmental Learning and Knowing Otherwise

Landing Bassene, Stephania Rovira Ochoa, Gohar Sharoyan, and Gabriele Budach

It is a bit paradoxical how these small objects can touch you that deeply, and represent such devastating realities. These are objects announcing real crisis.
—Nia, Team Meeting, November 2023

Introduction

Originating from different places in the Global North and South, our team includes Landing from Eloubalire, Gohar from Armenia, Nia from Columbia and Gabi from Germany. We all met in an MA programme (as students, graduates and teachers) at the University of Luxembourg. It is here at the heart of Europe that we learnt from Landing about the

L. Bassene • S. Rovira Ochoa • G. Sharoyan • G. Budach (✉)
Humanities, University of Luxembourg, Luxembourg City, Luxembourg
e-mail: langsenior@gmail.com; niarovira@hotmail.com; gohar.sharoyan@outlook.com; gabriele.budach@uni.lu

M. Matthews (ed.), *Ecologies in Practice and Learning: Arts Interventions in the Earth Crisis*, Palgrave Studies in Educational Futures, https://doi.org/10.1007/978-3-031-82338-1_2

situation in his home village and the threat posed to its people by rising sea water levels that have already destroyed half of the arable land.

Taking the reader through our journey, we discuss how stop motion animation enabled corporeal, embodied and empathetic learning, and how it helped us to overcome distance from a place and situation to which we—those not from Eloubalire—felt *distant*, shy and illegitimate to talk about, at first. We share how engaging in co-creating a stop motion movie created closeness to a place, a community and an environmental crisis that is happening far away from the place where we are writing this.

An Opening

Our text opens with an image. We see people, old and young, gathered in ceremonial festivity to celebrate the arrival of life jackets. The ceremony takes place in the village of Eloubalire in Southern Senegal. The image depicts a significant moment for the community. Symbolising more than the distribution of life-saving equipment, it marks a powerful manifestation of communal resilience and collective solidarity in the face of adversity. In what follows, we will share more about this community and uncover the different layers of meaning, folded into this event (Fig. 2.1).

Beginnings

Our story as a team started in Gabi's course on stop motion animation, a third semester MA course[1] at the University of Luxembourg, where students are invited to work on self-chosen topics and represent them in a short animation movie. The aim is for students to learn about a different tool and means of expression that explores the value of the visual and the material, going beyond the verbal mode, that still dominates ways of learning in this MA programme, like many other university courses in the humanities these days.

Landing was a student on the same MA programme; he made a movie on environmental destruction depicting the devastation in his home region in Senegal in the Global South, produced by industrialisation and capitalist overproduction in the North. He shared the film with the class. At this time, Landing explained to the group that the film and message dealt with the broader issue of environmental destruction, affecting

[1] Master in Learning and Communication in Multilingual and Multicultural Contexts.

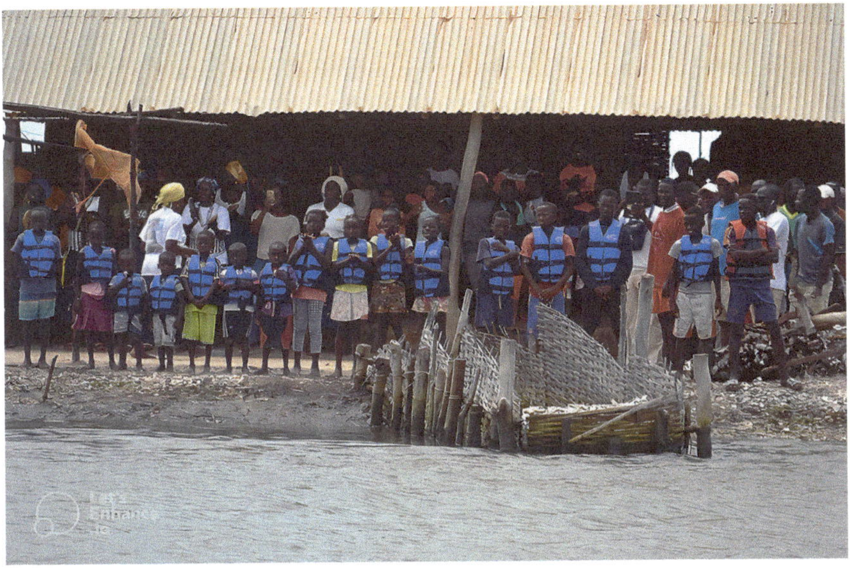

Fig. 2.1 Eloubalire children celebrating the arrival of life jackets

communities in his home region, but also on a global scale. It was only in the written reflection on his film making experience—assignment for this course—that he shared the story of his village giving more details about the alarming course of things. This is how Gabi who was teaching the course first learnt about Eloubalire.

She then asked Landing if he wanted to do another stop motion project—this time about the story of his village—and work together with Gohar and Nia, graduates of the same MA programme. Both were involved, at the time, in an outreach project that offered stop motion animation workshops to families from different backgrounds to raise awareness about issues of discrimination in Luxembourg. Landing agreed, and this is how our collaboration started. At the beginning, we (others) knew nothing about Eloubalire, as stories from southern parts of the world usually do not make it into media reports in Luxembourg, or central Europe more broadly.

Eloubalire: A Context

Eloubalire is an Island situated in the South of Senegal and near the Delta of the Casamance river. The community has about 1000 people who are Diola speakers. Traditionally, rice growing has been one of the major economic activities providing the community with its main food source. Due to rising sea water levels, arable land is flooded, and rice grains do not survive the increasing concentration of salt. As a result, agricultural land is being destroyed and devastated, and the community has lost about 50% of its arable land over the last 20 years.

On closer observation, there are media reports and academic research on such phenomena conducted in a number of world regions. Government programmes aim at implementing solutions that reach from the resettlement of populations (Chintam et al., 2021; Pinter, 2021) to improving local water management in places like India (Bobba, 2002), the Netherlands (Pauw et al., 2012) and the United States (Perry et al., 2022).

While wealthier parts of the world are in a position to propose viable solutions to affected communities, poorer communities are often left at their own devices. Protective measures had also been taken in Eloubalire: a dyke made from local soil was commissioned by the World Food Programme in 2002–2003, acting on governmental advice to be built by the locals. But, the dyke only lasted a few years and was soon destroyed by the incoming tide, a development that local residents could have predicted. This points to another problem, which is related to the fact that local situations often require local solutions, but that the local knowledge residing within communities is often not considered by policy makers and (non-)governmental agencies operating in an area.

For Eloubalire, Landing shared:

> Another problem is that solutions are proposed at state level, and sometimes they don't bother to see what local communities are going through. It's often said that anything done for me, without me, is automatically done against me. For example, if you want to carry out an activity in an area, the first thing to do is to check whether that activity has already been carried out there. And if it has, we need to find out how the local population developed it. (Team Meeting, January 2024)

Left to their own devices, the young generation, notably those living in the diaspora mobilised their forces and means to take their community into their own hands. A lot of learning took place for us—the other members of the team—during our collaboration. Although we all had heard about the problem of rising sea water levels as a global problem, we did not know or relate to any actual place or community where this was happening.

Our Creative Process and How Stop Motion Animation Affected Change for Us

Meeting at Dalmat

A first meeting took place in a café on campus. Landing shared photos from his village, many of which he had taken himself or found on the internet to explain about the situation in his community (Bassene, 2024). Through this source, we learnt first-hand about what life in Eloubalire was like and how the natural disaster had impacted the community. After the meeting, the rest of the group (Gohar, Nia and Gabi) were left quite sad and with many questions and doubts.

> We sat down … we spoke about the story, we all felt sad, we questioned how authorised we were to talk about this issue. (Gabi, Team Meeting, November 2023)

We had heard about a dramatic loss. But, having heard about it and having seen the pictures, we felt somewhat distant and shy. Was this our problem? Would it be right for us—not being Landing—to engage with the topic, and to work on something that was not ours, not connected to our own experience? Would it be right for us to speak on his behalf? Too often white people, or people from different places, are speaking on behalf of local (indigenous) communities to defend their interests, but in a sometimes patronising, uninformed manner serving, ultimately, only their own interests. What was our relationship with the people, the land, the 'problem'? Eloubalire being so far—and us feeling more detached than connected with what we saw to be a different, distant reality. Sensitive to the ethical concerns to be addressed in research, we were unsure if working together as a group was even possible.

Creating the Movie Together

Despite the doubt, we decided to give it a try and met to work on a stop motion animation together. We used small objects that we thought fitting for the purpose of the project, arranged them in front of the camera and started taking shots, one after the other. With the images Landing had shown us, in our heads, we imagined Eloubalire, with our hands, as if it was happening in a classroom in Luxembourg.

During the creation, what prevailed was a sense of play and fun, and a focus on what was to be represented in film:

> When we were making the movie, we had this playful mood, with all the colours and material. We almost felt guilty because we had so much fun … although it's about a really serious topic … we were trying to think how to represent the rice fields … then you [Landing] explained to us how the rice grows, the life cycle of rice … we learned that rice changes colour [close to the time of harvest] … we found a few stems that were not as green … we could integrate so many things, in the film, that we had learned. (Gabi, Team Meeting, November 2023)

> The process becomes very physical, and not so mental anymore. It's very haptic, very visual, it is really about the body, the object. (Gohar, Team Meeting, November 2023)

The time of filmic creation was not only one of playing with small objects, colours, shapes and arranging things for movement. It was also a site of learning, where Landing explained about the life cycle of rice, planting, growing and harvesting. We forgot about the sadness and discomfort we had felt during the meeting at the café.

> Trying to play with stuff means getting detached from sadness. (Gabi, Team Meeting, November 2023)

The film making felt like a deeply bodily, sensory experience of learning and quite different from only seeing the images—as we had done at the café. While the shared images spoke to us in a rational and emotional way, they did not enable us to overcome the distance we felt, geographically and experientially. We knew nothing about rice growing and what it meant to people in the community witnessing their main food supply being at risk and slowly disappearing. However, the handling of small objects and arranging them to be filmed seemed to allow for a different way of sensing and creating reality. Having fun while creating these arrangements together lifted our feelings of shyness and discomfort. It not only created a pathway for a different kind of understanding built on an embodied experience. It also gradually closed the gap and reduced the distance we had felt previously. Instead, familiarity, closeness and a new kind of bonding emerged based on an understanding that felt embodied and empathetic.

Watching the Movie Together

However, after watching the finished film, we sat in shock (Bassene et al., 2024). We were brought back to the dramatic reality we had been depicting in film, and felt deeply the seriousness and threat to life we just had documented in play (Fig. 2.2).

> Engaged in the process of the work, we didn't think so much about how sad it was. The story was virtual, and not reality yet … when the film was ready we could feel the sadness, … the silence that came after watching it … no one could express anything at that moment … you are kind of down to earth again, you are telling about real life, about something that is really happening. (Nia, Team Meeting, November 2023)

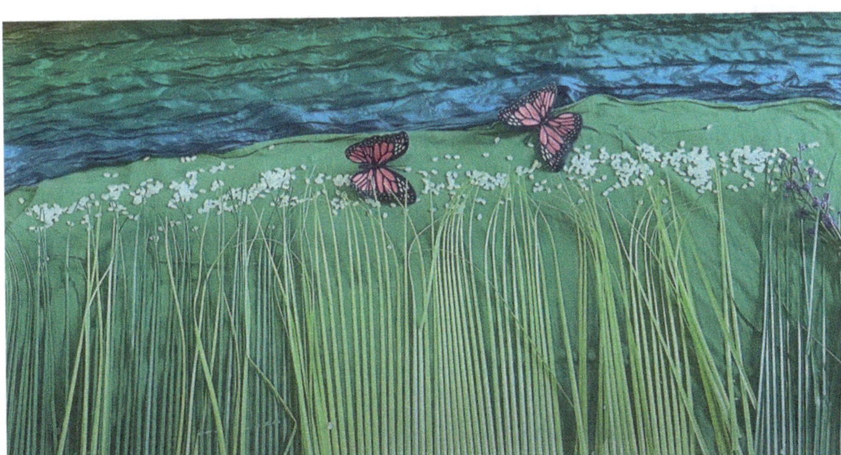

Fig. 2.2 Before the crisis (film scene from the authors' stop motion animation)

Eloubalire became REAL for us (not being from that place) through making a stop motion film. Assembling this new reality—while making choices about objects, music and speed—made the situation tangible and liveable, an embodied experience of a place to which we (except for Landing) had never been before.

> This is not one of those apocalyptical movies with huge waves rolling over a city … rather a destruction that comes slowly, gentle, without warning … but the results are dramatic. (Gabi, Team Meeting, November 2023)

> Thinking of the moments previous to seeing the sea levels rising and covering the rise fields … everything was beautiful and then came the destruction of that space. I think this was one of the most impactful moments for me … it made me aware of the devastation the territory has suffered … presenting it through material objects, was like the embodiment of awareness and allowed a deep process of reflection. (Nia, Team Meeting, January 2024)

Beyond this realisation and the embodied feeling of awareness, what also stayed with us was the new relationship of closeness we had developed during the film making, with Landing, with Eloubalire, with its people and their cause.

Becoming 'Border Thinking Eco-artists': Linking our Experience to Work on Decoloniality, Border Thinking and Eco-arts

While reflecting on our collaboration, we read literature related to concepts of decoloniality, border thinking and eco-arts. There were some concepts that resonated with us in particular. One notion that we connected with strongly was place, or more precisely the *place we are speaking from*. In our case, it makes a difference that we grew up in Senegal, Armenia, Columbia and Germany. Through this project, we recognised that there are multiple *loci of enunciation*, a notion proposed by Mignolo (2009) that suggests that there is not one (total, all encompassing) view and vision of the world but a multitude of perspectives that are grounded geographically and experientially within the position enunciated by different speakers. For us, this implied that we not only grew up with different ways of approaching, seeing and understanding the world, but that we also embody these different views.

Thus, we align with concepts of decoloniality that contest the idea of universality as defined by Western/Northern epistemes and their (implicit) claim of superiority and hegemony. This is echoed in what Vazquez (2014) says:

> [Decoloniality] is in contrast with thinking centered in the Western philosophical tradition, in which modernity in its different facets is assumed to be the totality of reality. For decolonial thinking [...] there is an outside, something beyond modernity, because there are ways of relating to the world that come from other geo-genealogies, non-Western and non-modern. (Vazquez, 2014, p. 173)

Another dimension of this discussion we felt deeply connected with was the importance of the/our body. As Cetshwayo Mabhena (2019) says:

> The *body* is a vivid site of power of politics, a locus of enunciation from which we feel, see, think and speak. ... Just like people and their bodies, ideas, thoughts and knowledge have a biography that traces their genealogy and provenances. (online source)

The *body* (in plural) stands here for us—four human individuals—aware of the dividing lines of discourses on modernity and coloniality, trying to find a common ground to work together, respectfully and being ethically responsible. But, how to engage with the legacy of Western modernity and colonialism, past and present, in our group? How to acknowledge and deal with the fact that we as individuals speak from a place (locus of enunciation) that sits in our bodies, very materially? Not addressing it would mean to implicitly reproduce Western hegemonic discourses. As Rosalba Icaza says:

> It is from this *placeless/bodyless* position that the histories of certain bodies [...], places [...], events or memories [...] [become seen] as the 'normal' ones and as universalised and reproduced as 'common' senses from which we think the international and the global. (Icaza, 2017, p. 8)

Hence, we realised that we had to engage in an exercise of 'border thinking' that included our (own) bodies.

An Exercise in 'Border Thinking' Bringing Together Place, the Body and Stop Motion Animation

We took a bet trying to work with 'stop motion animation' to see if it could serve as a site of productive 'border thinking' for us. Having done years of work on this approach (Budach & Sharoyan, 2020; Budach et al., 2021, 2022), there was hope it could produce a new and different *locus of enunciation*, different from the ones we were coming from. Would we be able to experience 'border thinking' as something constructive? Of which Cetshwayo Mabhena (2019) says:

> Border thinking is liberation thinking. It gives us a locus of enunciation that is above and beyond ourselves. (online source)

However, there were two conditions needed for it to work. To make our stop motion project the deeply connecting experience of 'border thinking' it became, we needed Landing to share his local knowledge with us. Landing showed us that knowing sits in bodies—his body—territories and its local histories in contrast to disembodied, abstract, universalist knowledge that generates global designs (Mignolo, 2009, 2010). As described by Rose (1997, p. 308), he helped us recognise that knowledge is situated and needs to be seen 'from specific locations' that are 'embodied and particular, and never innocent' (Diniz de Figueiredo & Martinez, 2021).

The second condition was to create an alternative place of encounter and enunciation which would help us with overcoming the doubts and division we felt at the start of the project; a place of border thinking that allowed for 'a physical sensual experience of a self-in-between that is a plural self', as Rosalba Icaza (2017, p. 35) says. We discovered that Eco Art Education as a conceptual framework, and stop motion animation as the material site of an eco-artistic practice, provided much of the kind of space we were needing.

Eco-visualisation Bringing Together Artistic Practice and the Concept of *Locus of Enunciation*

Thinking about our experience, we see it as a practice of place-based education (Sobel, 2004) that is critical and situates learning within local communities and the environment (Bertling, 2013, 2015). For us, recreating

Eloubalire through stop motion animation became a form of eco-visualisation and exercise in ecological imagination of which Dean and Bertling (2020) say that it can engage students to empathise with more-than-human others, to expand their consciousness for other living things and to envision more ecological ways of being (Bertling, 2013). Since we could not be immersed in the place of concern and go to Eloubalire ourselves, we needed to visualise and imagine it. For this, it took a person FROM that place, able to share the LOCAL knowledge and *place of enunciation as embodied knowledge* with us:

> If Landing wasn't part of this, we wouldn't have had this first-hand information, we would be writing about an environmental disaster as a general thing. We know, we hear about global warming. We hear about raising levels of waters and everything. But his presence as someone from that place ... the message was so clear, and it brought us so close. All the information you told us about the life jackets, the rice field, the dyke and everything I feel like this is very specific knowledge and it's coming really from that place. (Gohar, Team Meeting, November 2023)

Landing has been our teacher, and we could not have done this project without him. Sharing his experience with us, we experienced a place deeply imbued with local knowledge and culture, and at the same time entangled with national and global politics. All these dimensions were built into in our work, which we see as an exercise of 'border thinking through eco-visualisation'. We agree with Dean and Bertling (2020) that such an exercise provides students with a critical awareness of their own location as a historically situated site of enunciation. It also offers the option of relating to the world as a 'place of different words of meaning' (7), that is there to be explored rather than consumed.

While we learnt about a specific place and its environmental disaster, we also discussed our relationship with nature, more generally, and shared that:

> Everything is in nature, and going back to nature doesn't necessarily mean to go and sit in a forest. ... We are nature and often much closer to nature and the environment [than we think we are]. (Gohar, Team Meeting, November 2023)

Stop motion animation, the close and detailed engagement with small objects and imagining the story of Eloubalire shared with us by Landing helped us close the gap between the human-and-more-than-human divide

(Barad, 2007; Braidotti, 2013). It made us feel like artists and opened a new dimension of embodied awareness, for us and for others.

> I saw this in our stop motion … based on the story I told, we were able to use our artistic imagination to choose colours, mosaics and a whole lot of objects that enabled us to create a work of art … that's an exceptional thing. (Landing)

Feeling like eco-artists, in our own (individual) capacity, we felt empowered and remembered Hilary Inwood (2008, p. 59) who says: 'Eco-artists can be thought of as midwives for the earth'. We feel particularly strongly about the point she raises on eco-art's 'ability to stimulate learners' minds, yet also touch their hearts [which] makes art education a powerful ally in fostering ecological literacy' (59). It is through engaging in artistic practice and learning about one's own ability of artistic expression that a new way of finding one's voice opens. This voice can counter or be alternative to hegemonic ones, such as science discourse which heavily dominated environmental education in the past.

> When I tried to find texts about rice growing and rising sea water levels they came mostly from science … taking a scientific stance and way of presenting results, describing the measurements they did. No word about why there is climate change … it really felt quite disembodied and unrelated to the people who caused it, or who are suffering the consequences … I think the humanities have a responsibility to change the discourse and add a human dimension. (Gabi, Team Meeting, February 2024)

We agree that eco-visualisations can have a powerful impact on both, their makers and viewers. They can become new forms of data visualisation (Taylor, 2017) and inform people, as Dean and Bertling (2020) say; or incite them to 'reflect and inspire deep understanding and engagement' (2020, p. 60). They therefore deserve an important place in the (arts) curriculum and can potentially help educate students towards building a new relationship of appreciation and connection with the planet.

Celebrating Life: Creating 'Knowing Otherwise'

We would like to thank Landing who has been our teacher on this journey. We also would like to acknowledge the Senegalese diaspora, young people from the southern region of Senegal, Eloubalire and beyond, who are making their lives abroad, but have set out to make their home village and region back home safe, deploying enormous effort, human and financial, to achieve this ambitious goal. They are collecting funds to build a road connecting the home village to the main road in the region, which will then make it possible to build a stone dyke that will be sourced and brought from other parts of Senegal.

We are grateful for what we have been able to learn together, through *listening* to Landing, *seeing* his photographs taken year after year for the last two decades and documenting the incremental, but steady loss of land; and *doing* this stop motion film together. Our dialogue has been rich and invited *voices* and *knowledge* that did not have a place in our curriculum before. And, this is not the end of the road: learning, teaching and creating with one another in this way allowed us to bridge boundaries of place, experience, history and responsibility and helped us feel more connected and closer to Eloubalire, advocate for its cause and feel eager 'to do something'. There is hope that arts-based work of this kind enables us to travel, share, feel and understand, even while speaking from different places.

> We didn't travel physically, but it's like we travelled with our bodies, using objects and feeling with our senses—the high tide rolling in. (Nia, Team meeting February 2024)

Revisiting the Arrival of the Life Jackets

Why do we start and end this chapter with the image of the life jackets? Because it says something important about the people of Eloubalire and the way they see life and the world. The image depicts the arrival of 104 life jackets and 6 two-way waterproof radios that were bought by the Eloubalire diaspora with funds raised from a call for donations, after the drowning of eight members of the community in a boat accident. The arrival of the life jackets—proudly worn by all members of the community, young and old—is greeted with a festive ceremony. Women dance and show that even in the eyes of death and disaster, there is no giving in to

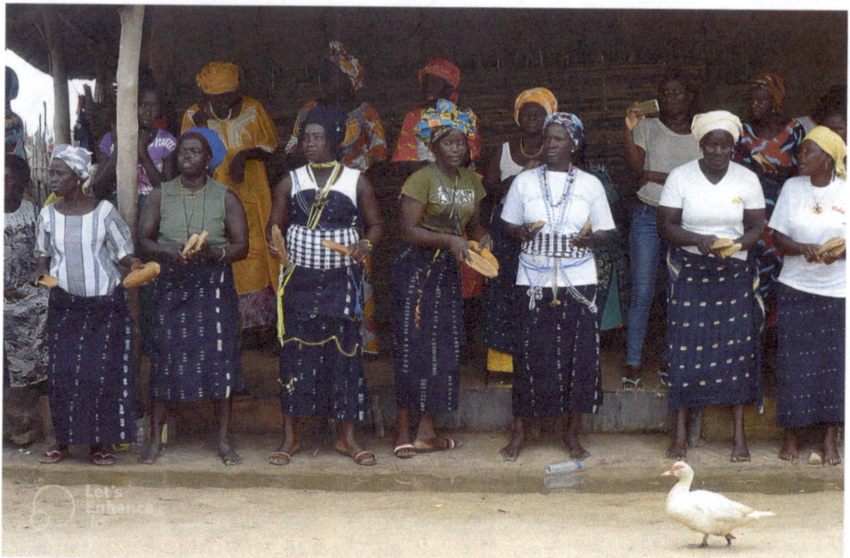

Fig. 2.3 Women dancing at the ceremony for receiving the life jackets

sadness, adversity and desperation. While the dead are remembered, joy, positivity and a hopeful looking ahead prevail. It says that there is the need to act for change and we are ready (Fig. 2.3).

Conflict of Interests (4) The authors have no conflicts of interest to declare that are relevant to the content of this chapter.

References

Barad, K. (2007). *Meeting the universe halfway: Quantum physics and the entanglement of matter and meaning*. Duke University Press.

Bassene, L. (2024). *What brings us together: A shared concern: From Eloubalire, Landing's story*. Retrieved August 12, from https://www.youtube.com/watch?v=nK6UNc0LOc8&t=2s

Bassene, L., Rovira Ochoa, S., Sharoyan, G., & Budach, G. (2024). *Eloubalire [stop motion animation]*. Retrieved August 12, 2024, from https://www.youtube.com/watch?v=c7Oxt8gH6P4&t=

Bertling, J. G. (2013). Exercising the ecological imagination: Representing the future of place. *Art Education, 66*(1), 33–39.

Bertling, J. G. (2015). The art of empathy: A mixed methods case study of a critical place-based art education program. *International Journal of Education & the Arts, 16*(13). Retrieved August 12, 2024, from http://www.ijea.org/v16n13/

Bobba, A. G. (2002). Numerical modelling of salt-water intrusion due to human activities and sea-level change in the Godavari Delta, India. *Hydrological Sciences Journal, 47*(Sup1), S67–S80.

Braidotti, R. (2013). *The posthuman*. Polity.

Budach, G., Efremov, D., Loghin, D., & Sharoyan, G. (2021). Conversing with humans and objects: On repetition and the curative power in animation making. *The Canadian Modern Language Review, 77*(4), 353–373.

Budach, G., & Sharoyan, G. (2020). Exploring 'vibrant matter' in animation making. *Language and Intercultural Communication, 20*(5), 464–481.

Budach, G., Sharoyan, G., & Loghin, D. (2022). 'Animating objects': Co-creation in digital story making between planning and play. In V. Lytra, C. Ros i Solé, V. Macleroy, & J. Anderson (Eds.), *Liberating language education: Personal, aesthetic and political perspectives* (pp. 277–296). Multilingual Matters.

Chintam, K., Jackson, C. T., Dunn, F. B., Hall, C. A., Nathan, S., S. I. N. D. H. U, & Navarro-Serer, B. (2021). A concerted and equitable approach to managed retreat. *Issues in Science and Technology, 37*(4), 70–73.

Dean, K. P., & Bertling, J. G. (2020). Eco-visualizations: Facilitating ecological relationships and raising environmental awareness. *Art Education, 73*(3), 54–61.

Diniz De Figueiredo, E. H., & Martinez, J. (2021). The locus of enunciation as a way to confront epistemological racism and decolonize scholarly knowledge. *Applied Linguistics, 42*(2), 355–359.

Icaza, R. (2017). *Border thinking and vulnerability as a knowing otherwise*. Retrieved March 7, 2024, from https://www.e-ir.info/2017/06/09/border-thinking-and-vulnerability-as-a-knowing-otherwise/

Inwood, H. (2008). Mapping eco-art education. *Canadian Review of Art Education: Research and Issues, 35*, 57–73.

Mabhena, C. (2019). *On the locus of enunciation*. Retrieved March 7, 2024, from https://www.sundaynews.co.zw/on-the-locus-of-enunciation/

Mignolo, W. (2009). Epistemic disobedience, independent thought and decolonial freedom. *Theory, Culture & Society, 26*(7–8), 159–181.

Mignolo, W. (2010). Delinking: The rhetoric of modernity, the logic of coloniality and the grammar of decoloniality. In W. Mignolo & A. Escobar (Eds.), *Globalization and the de-colonial option*. Routledge.

Pauw, P., De Louw, P. G., & Essink, G. O. (2012). Groundwater salinisation in the Wadden Sea area of The Netherlands: Quantifying the effects of climate change, sea-level rise and anthropogenic interferences. *Netherlands Journal of Geosciences, 91*(3), 373–383.

Perry, D. C., Ferguson, W., & Thornber, C. S. (2022). Salt marsh climate change adaptation: Using runnels to adapt to accelerating sea level rise within a drowning New England salt marsh. *Restoration Ecology, 30*(1), e13466.

Pinter, N. (2021). True stories of managed retreat from rising waters: As more communities face threats from rising sea levels and climate-induced flooding, a rich 140-year history of relocation projects offers valuable insights. *Issues in Science and Technology, 37*(4), 65–74.

Rose, G. (1997). Situating knowledges: Positionality, reflexivities and other tactics. *Progress in Human Geography, 21*, 305–320.

Sobel, D. (2004). *Place-based education: Connecting classrooms and communities.* The Orion Society.

Taylor, P. (2017). Artistic data visualization and assessment in art education. *Visual Arts Research, 43*(1), 59–75.

Vazquez, R. (2014). Colonialidad y relacionalidad. In M. E. Borsani & P. Quintero (Eds.), *Los desafíos decoloniales de nuestros días: Pensar en colectivo* (pp. 173–196). EDUCO/Universidad Nacional de Comahue.

CHAPTER 3

Compound 13 Lab, Mumbai: Learning Through Waste Ecologies

Graham Jeffery, Ben Parry, and Sharmila Samant

Material Waste Flows in Mumbai

Mumbai is a major financial, commercial, and entertainment hub. It is also one of the most unequal cities in the world, where extreme poverty co-exists with extreme wealth. In a mega-city of 21 million people operating without centralised infrastructures to match the scale of the population, waste flows through a complex network of brokerage and trade in which the municipal authorities play only a partial role. Due to limited municipal waste collection and household segregation, domestic and commercial waste, made up of wet waste (food, organic and biodegradable stuff) and

G. Jeffery (✉)
Business and Creative Industries, University of the West of Scotland, Paisley, UK
e-mail: graham.jeffery@uws.ac.uk

B. Parry
School of Art, Film and Media, Bath Spa University, Bath, UK
e-mail: b.parry@bathspa.ac.uk

S. Samant
Compound 13 Lab, Mumbai, India

M. Matthews (ed.), *Ecologies in Practice and Learning: Arts Interventions in the Earth Crisis*, Palgrave Studies in Educational Futures, https://doi.org/10.1007/978-3-031-82338-1_3

some hard waste like plastics and packaging material, finds its way to landfill where it is scoured through by waste pickers to extract recyclable materials. However, most of the city's hard waste escapes landfill, through processes where it re-acquires value, traded via a complex web of social and material relations. Discarded objects are traded in an intricate supply chain of collectors, brokers, and dealers, before they are broken down, repurposed, or recycled (Doron & Jeffrey, 2018). Dharavi, home to 1.4 million people on 2.1 square km of prime land in the centre of Mumbai, is one of the largest informal settlements in the world. Hundreds of workers sort and trade plastics and other recyclable materials in an area of Dharavi known as 13 Compounds.

As public awareness of the importance of recycling has grown, the term has been adopted by frontline workers as a claim to the broader societal value of their work. Vinod Shetty, founding director of ACORN India and a labour lawyer in the Bombay High Court, places emphasis on human rights and labour rights, attempting to shift prevailing attitudes about those handling waste as dirty, undeserving, and disposable, and instead acknowledging them as 'green collar workers' (Jeffery & Parry, 2023), repositioning their work as valuable and environmentally significant.

These 'human infrastructures' (Jeffery & Parry, 2023) of the informal waste recovery industry considerably reduce the volume of plastic that goes to incineration or landfill. The alternative trade networks of informal recycling create supply chains that contribute to the wider economy as well as sustaining hundreds of thousands of family livelihoods. Dharavi is now a key target for speculative redevelopment: the current version of the Dharavi Redevelopment Plan envisages demolishing most of the existing structures, replacing the piecemeal development with a high-rise business district which will be entirely unaffordable for those currently living and working there.

India has achieved one of the highest recycling rates in the world, developed through grass roots entrepreneurialism. The informal waste trade has allowed many of Mumbai's poorest inhabitants to find work; some have developed their own businesses over generations. Unlike today's institutional approach to sustainability, the origins of the informal plastic recycling industry did not come from the lofty aspiration to generate a 'circular economy'; rather, it arose from basic needs, livelihood, and frugality. The 'waste commons' of available raw material, discarded and scattered across public spaces, can be mined for value (O'Hare, 2022).

People undertaking this work generally have very limited choices, often with family and caste/class ties to this occupation, and costs of entry are low (Gill, 2010; Gidwani, 2015). At its peak, we estimate that Dharavi's 13 Compounds recycled 80% of Mumbai's hard waste, supporting the livelihoods of around 250,000 informal waste workers, ranging from waste pickers scouring the streets, to *kabadiwalas* (neighbourhood waste collectors), *feriwalas* (transporters) sorters, brokers, dealers, and remanufacturing. The 13 Compounds is just a small part of Dharavi's extraordinary economic success: as one of the world's most famous informal settlements, it has resisted numerous attempts at wholesale redevelopment over recent decades (Fig. 3.1).

Despite its remarkable achievement in building a highly efficient self-organised circular economy, Dharavi is far from an ecological paradise. It does however contain many other examples of urban innovation in design and infrastructure at neighbourhood level. We have referred to it elsewhere as a 'smart city from below' (Jeffery & Parry, 2023). Places like Dharavi tend to be rendered as junk or waste land, ripe for speculative 'improvement' by the state and by the powerful, because of the negative associations of being labelled as a 'slum': a legal term with origins in colonial governance (Chhabria, 2019). By extension, people who live in such

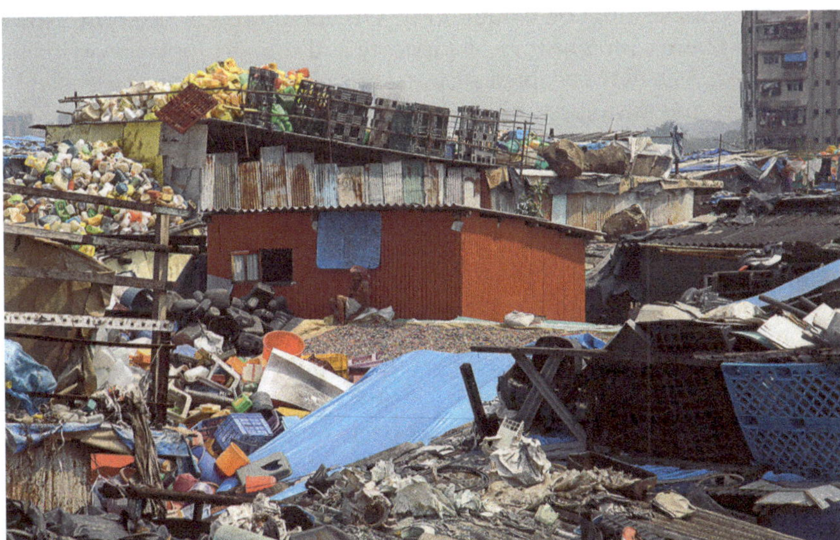

Fig. 3.1 Rooftops of the 13 Compounds, Dharavi, Mumbai, 2012. (Image: Ben Parry)

settlements are treated as disposable, as if nothing can be learned from them and they do not matter. These forms of exclusion are partially rooted in India's caste/class system. Families with intergenerational knowledge of waste work are simply undertaking their 'assigned roles', cleaning up the rubbish and waste generated by the upper castes (Doron & Jeffrey, 2018). Waste work therefore becomes synonymous with the differential valuing of human life and with untouchability and disposability.

Creative Learning and the Waste Economy

As a hybrid of urban lab, art space, community centre, maker space, and classroom, the Compound 13 Lab project weaves together pedagogical experimentation, arts practice, ecological concerns, and social solidarity in conversation with the everyday realities and life challenges of people living in Dharavi. Its name, *Compound Tera* (13 Compounds) in Hindi, makes a connection between the 13 separately owned compounds that originally formed Dharavi's recycling district, as well as the idea of combining or compounding different materials, perspectives, and experiences.

The project, which began in 2012, is built on a collaboration with ACORN (Association of Community Organisations for Reform Now) India, who have a long history of advocacy and support for waste workers. At that time ACORN occupied a former recycling workshop (*gowdown*) in the heart of the waste processing and recycling activity, with membership of over 500 families living and working within Dharavi and the nearby Bandra slum. In 2011, a series of state-led expulsions forced many ACORN members to relocate following the demolition of 450 homes and small recycling businesses along the steel water pipelines that run above ground on their journey to carry the city's fresh water supply (Parry, 2015). Many are familiar with this iconic pipeline location through various forms of film and media including scenes from Danny Boyle's *Slumdog Millionaire* (2008) to articles in National Geographic (2007), and Kevin McCloud's Channel 4 documentary, *Slumming It* (2010).

The initial remit of ACORN had been to improve the welfare and working conditions of waste worker communities across the city. Over the years supporting these workers, ACORN increasingly turned their focus to the children and young people of families making their living from the city's waste, attempting to change their life trajectories and free them from the hardships and toxic burdens of waste work. Central to this approach is the imaginative use of arts-based methods to encourage broad participation, alongside more traditional classes in maths and literacy. Compound 13 Lab augments this work by inviting artists from a variety of backgrounds, including visual art, theatre, music, and craft, to work with ACORN's young people, providing a platform which supports an ambitious programme of creative engagement. The principal vehicle for this is an extended artist residency programme, which also sets out to support early career social practice artists.

The activities generated by the residency programme help establish an innovative and alternative approach to the 'maker space' and 'classroom', providing free learning resources for young people aged 8–25 who would otherwise have few opportunities of this type. Through a process of increasingly self-organised making and learning, Compound 13 Lab offers ACORN's young people access to current technologies in digital media, music, design, 3D printing, and digital fabrication.

One key concern was how artists might work more ethically and effectively in such contested spaces. Dharavi remains a site of extreme poverty and hardship. The well-trodden issues of artists working in deprived communities (Kelly & Jeffery, 2016) are further problematised by the international fame of Dharavi as an urban phenomenon of bottom-up ingenuity, as well as a highly visible site of socio-material possibility that continues to attract artists. Projects which succumbed to the lure of art world recognition by inviting Mumbai's cultural elite into Dharavi's neighbourhoods to 'look at art' felt very out of context, falling into a similar category of outsider presence as the popular slum tours that take tourists on a journey through Dharavi's contrasting neighbourhoods. The Lab therefore set out to promote more meaningful and reciprocal interaction between visiting artists and community members, beyond the short term 'encounters' that are more common in community arts projects. Inviting young artists that had trained on Shiv Nadar University's MFA programme provided challenging but invaluable professional practice experience.

Within academic and urban research, Dharavi has long been an international 'go-to' site for all manner of excursions, projects, and case studies: a contender for the most documented informal settlement on the planet and a hotspot for media representation (Jeffery & Parry, 2023). We observed that the work being produced by NGOs, planners, and academics was being published internationally but was almost never returned to community members. An early proposal was the creation of Dharavi Digital Public Library, where residents could access materials and texts being produced about their everyday lives and the possible future of their communities (Parry, 2015). Another important influence was the idea of establishing a 'maker space', building on the intense making and repair culture within the neighbourhood. Within the community, there are deep knowledges of manufacturing and repair, but few spaces or opportunities for people to experiment, or take time to think, create together or innovate. From the outset, we asked: what kind of spaces could be created where knowledges could be exchanged within and beyond the community?

Learning Through Neighbourhood Practices

We set out to connect the Lab's work into wider international conversations about the future of learning and issues of urban sustainability. Alongside the creative programme, the Lab team conducted in-depth research and data collection, quantifying and gathering data on the quantities of plastic that the informal sector takes out of the system. Dharavi's 13 Compounds, as one of about 20 such sorting and trading hubs across the Mumbai Metropolitan Region, alone handles around between 500 and 800 tonnes of plastic waste per day, around 187,000 tonnes annually (Pancholi & Kamble, 2023). Effectively, India's informal waste sector provides a waste management service operated by the urban poor without any funding or acknowledgement by the state, substantially subsidising the city as a whole, which balances its enormous waste problem on the back of the poorly remunerated labour of the urban underclasses. As a 'thinking and making' space linked to international research networks, C13 Lab set out to connect the hidden knowledges embedded in waste work, effectively local cultures of innovation, with broader international debates about sustainable futures. It attempted to reframe public debate: representing waste work and the marginalised people undertaking it as performing valuable infrastructural maintenance work.

'Sustainable urbanism' models are widely celebrated (Joshi et al., 2011) but Dharavi is rarely represented as a set of intergenerational practices and knowledges that could offer frameworks for 'urban learning': considering how urban sustainability is created through the lived practices of the neighbourhood. Perspectives of urban development and patterns of urban inhabitation informed work undertaken in the Lab, particularly the materials, resources, and knowledges involved in the everyday life experiences of its users. The Lab's data gathering work included a mapping and photography project in which young participants developed a typology of waste work actors and processing centres by taking geo-tagged photographs using smartphones. These were later uploaded to an interactive map showing key locations and waste flows, directly contributing to our research mapping the movement of waste materials through Dharavi's trade networks.

Artist Residency as Method

Creative and cultural learning enables people to express ideas, feelings, and aspirations about their neighbourhoods, including through creative action (Jeffery, 2005; Leeson, 2017). Some distinctive skills that artists and creative producers bring that add value and complement the expertise of other researchers, facilitators, and educators include encouraging encounter and interaction between diverse people; animating shared and public spaces; infusing meaning, exploring how a place feels and is experienced; expressing the intangible and the invisible; and stimulating new ideas and approaches through creating situations in which people encounter new or different ideas.

Ongoing activities at the lab include art and design classes, workshops, training sessions for members aged 8–25, talks, film screenings, and an ongoing programme of guest residencies, internationally and within India. Young people work alongside artists on a variety of different projects, which advance through a process of co-creation, shared-learning, and co-authorship. These activities are broadly structured around a set of themes which weave together practices of participatory action research, knowledge production and learning, creative ideation, and experimental making (Jeffery & Parry, 2023; Wakeford & Sanchez Rodriquez, 2018).

One of the first and positive interventions of the Lab was to use the artist-residency platform to address gender divisions that meant that some of ACORN's activities tended towards all-male groups. Given the prevailing climate of increasing communal tension within India (Banaji, 2024) the achievements of ACORN in promoting equality and integration should not be underestimated. The Lab consciously explored how approaches to participatory art projects and their processes of engagement can challenge them, creating a safer space for young men and women to mix, including Hindu and Muslim youth from different backgrounds.

One such intervention involved artist Aqui Thami, who with her partner Himanshu runs a parallel initiative, Dharavi Art Room, a longstanding creative space for young people to make art and learn. Thami provides a full account of her project, 'Precious' in *Waste Work: the art of survival in*

Dharavi (Jeffery & Parry, 2023). She initiated intergenerational conversations between young women and their mothers and grandmothers about jewellery which for various reasons had been lost or sold, usually to do with financial hardship. These narrated stories were recorded, and pencil drawings of the lost jewellery were turned into 3D digital models, which were printed and placed into jewellery cases. Recordings of the stories of each item were included in the cases. One case was returned to the family and the other was exhibited, as a means of sharing public dialogues about precious items and the ways in which different notions of value had been embedded in these 'lost' objects (Jeffery & Parry, 2023).

Issues of access, safety, and violence against women in the context of very male-dominated workspaces were rarely discussed publicly. Traditional gender roles within family life and the expected share of domestic labour meant young women were discouraged from attending spaces like ACORN or the Lab. Bringing more gender awareness and equality into the work of the lab required foregrounding women's voices and experiences. Given how internalised patriarchal structures are, this required a slow, subtle process of trying to unpick male dominance in the lab space—involving learning and unlearning, trying things out, and experimenting, with facilitators observing, listening, and encouraging voice. These informal values of the lab are important because they too provide 'human infrastructure': facilitators model mentorship, and practices of care, respect, and dignity, which are not always explicitly articulated but are implicit in expectations about how people relate to each other. Lab participants undertake journeys across the city to visit unfamiliar spaces such as museums and concert halls, making such spaces less intimidating and building knowledge and confidence.

Combining traditional drawing with digital design capabilities, Thami's 'Precious' project exemplified how artists can make use of 3D modelling and printing technology. The practical processes and material relations of 3D printing and plastic recycling have a specific connection to the political ecologies of waste and Dharavi's urban landscape. Considering alternative futures for plastic, drawing upon open-source resources, was intended to expand the horizon of Dharavi's self-built urbanism; empowering young people to get involved in design and manufacturing processes. The affordances of rapid prototyping technologies were easily comprehensible for participants who had grown up around the plastics industries, familiar with commercial plastic manufacturing. Within Dharavi's urban fabric, integrated manufacturing technologies are highly visible, at street and neighbourhood level, rather than kept hidden in closed-off industrial areas. One of our objectives had been to find out if we could use Dharavi's plastic waste to manufacture recycled filament, but as technologies improved with faster print heads and finer nozzles the demand for higher quality and more stable filament increased. Even so, the use of 3D printers proved challenging, with the requirement to master 3D design software, as well the need for appropriate technical support. The environment itself, high temperature fluctuation, humidity, power outages, and use of fans, all worked against the regulated and stabilised conditions necessary for reliable printing with plastic filament.

British artist Ian Dawson undertook several residencies that used plastic and 3D printing as a tool of cultural and creative exchange (Fig. 3.2), working with young people to demonstrate the possibilities of the technologies, from 3D scanning of objects and environments, to assembling digital artefacts as well as printing objects to take away (Jeffery & Parry, 2023). Amongst his many collaborations with the Lab, Dawson brought in a project with London's Horniman Museum: objects from the

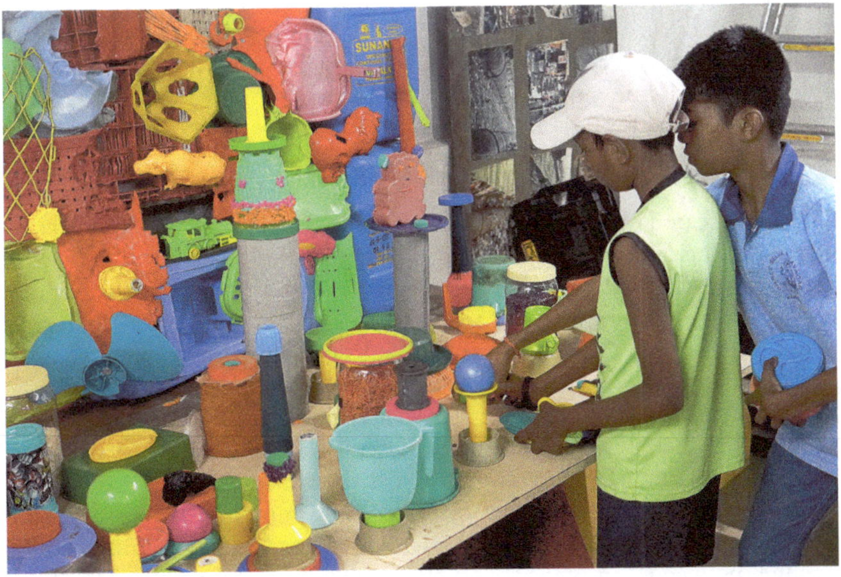

Fig. 3.2 Plastic sculpture workshop led by Ian Dawson at Compound 13 Lab, Dharavi. (Image: Ben Parry)

Horniman's extensive collection of Indian artefacts were shared with community members. Each selected one item which corresponded to familial objects with value in their own community. These artefacts were then scanned and reprinted and the plastic replicas were then returned to participants.

Finding different ways to record, photograph, and document what was happening within the neighbourhood raised questions of citizen journalism. Amol Lalzare, a former rickshaw driver, developed his own journalistic practice and facilitated participants' storytelling skills through a series of media workshops: in these forms of learning exchange, projects are designed with learning and development for everyone in mind (Jeffery, 2005). Lalzare's films are a potent example of how participatory approaches to storytelling using readily available media tools can enable the voices of the urban poor to be amplified. A participant led, issue-driven approach can efficiently develop citizen research skills, journalism, and filmmaking using the full potential of the mobile phone and free-to-use video editing software.

Only a minority of people in Dharavi have formal education beyond basic numeracy and literacy, and few can afford to attend further or higher education, so learning and skills acquisition is undertaken via other means, including kinship networks, and increasingly through digital and online engagement via the rapidly growing use of smartphones.

Global Conversations on Climate Action

The Lab created opportunities to incorporate community voices into creative learning and exchange through a variety of different methods, with the everyday struggles and stories of participants shaping the themes and content of its activities. On occasion our work was internationally recognised, and we were invited to contribute to projects elsewhere.

For example, Compound 13 Lab was invited to partner with the Museum for the United Nations 'Global We' project in what is the largest recorded global climate dialogue to date, involving 10,000 participants. Connecting people in 22 locations across the world via a network of portals (recycled shipping containers with real-time immersive video-streaming technology) 'Global We' facilitated over 3000 conversations about climate change between remote sites, without the need to travel (UN Live, Museum for the United Nations, n.d.). Between September 2022 to April 2023, the Mumbai portal, placed in the grounds of a local Dharavi school, participated in over 120 international conversations. Inviting diverse people from across the city, our participants engaged in dialogues, workshops, rap battles, and music sessions with people in locations including Jakarta, Kigali, Johannesburg, Mexico City, and Dhaka.

During the same period ACORN and the Lab had begun to engage a women's waste picker collective that had, unusually, organised themselves to protect against the dangers of waste collecting, including harassment and territorial poaching. Instead of being subjected to daily-wage fluctuations dependent entirely on the volume and kinds of waste gathered on the streets of Dharavi during a single shift, they pooled their catch and distributed earnings equally among the group. Having incorporated their stories into film, the UN Live portal enabled us to amplify their normally unheard voices and lived experiences to a wide international audience, including policymakers and stakeholders of the United Nations Climate Change Conferences in Egypt and the UAE.

Alongside schoolteachers, pupils, environmentalists, activists, artists, scientists, and citizen scientists, the contributions of waste workers and the network of Dharavi hip-hop artists connected the ambitions of the Lab with global conversations about sustainability and climate change. Bringing the women's waste picker collective into the portal, foregrounding their voices and perspectives as frontline workers in the informal waste industry, and enabling conversations with, for example, indigenous youth activists in Alaska, offer rare and significant exchanges that move beyond the confines of the portal into the orbit of decision makers.

Building Links with Artisanal Cultures of Making

Building up a programme of classes and workshops, responding to everyday conversations also deepened our understanding of the demographics of Compound 13 Lab's youth and the families they belong to. Many of them are from families of craft workers belonging to various *Vishwakarma* castes. Their identity is strongly tied to their profession. From various parts of the country, their traditional expertise lies in being carpenters, stonemasons, blacksmiths, bronze/copper smiths, or goldsmiths. Migration to Dharavi was often linked to the lack of local opportunities for practising their family's previous occupation. Knowledge and skills are thus less likely to be passed on intergenerationally, as would have happened in their hometowns.

Dharavi's young people identified themselves as insiders and the visiting artists/designers/filmmakers as outsiders: they were highly conscious of the differences of formal education, class or caste, and language. We felt that it was essential to bring artists/craftspeople from different backgrounds to interact at the Lab, to make interactions more fluid and provide role models for young people which did not simply reproduce elite kinds of education. Crafts which used recycling and locally available material were chosen. We hoped to challenge hierarchies, such as the downgrading of traditional crafts within formally taught art and design since colonial times, which have inflected discourses surrounding contemporary art, and social, political, and popular culture.

A significant example that used traditional craft to tell the story of recycling in Dharavi took place during a residency by Mamoni Chitrakar, an eighth-generation *patua* (painter) from Medinapore in West Bengal. Mamoni was trained as child by her mother, Swarna Chitrakar, in the art of *Patachitra* and both their work has been in exhibitions across India and abroad. *Patachitra* is a traditional form of storytelling from Bengal: as village bards they travel around neighbouring villages singing and painting on various themes ranging from mythological stories to historical as well as contemporary accounts, mixing them up with socio-political issues. The stories are first written then composed into song (*patar gaan*) and the rhythm of the words provides the structure to the painting. The painting is made on scrolls of paper unfolding either vertically or horizontally, with bold colours, lines, and strokes using natural colours that are made by hand using local plants. As the nine-metre long scroll unfolds, Mamoni sings the *patar gaan*. Having painted other river stories earlier, the Mithi river scroll became unique as it recounted the history of the waste pickers and recyclers in Dharavi (Fig. 3.3).

During the workshops Mamoni walked around the nature park pointing out the trees and plants from which she makes her colours. Demonstrating how a specific colour can be made from herbs and condiments available in everyone's kitchens or readily available local plants was inspirational for participants, who would later see themselves in the stories being told. Used coconut shells were best for mixing, and how to store colours was another important discovery. Paints are quite expensive and young people generally are not given colours due to paucity of means at home: acrylic colours provided at the Lab generally dry up very quickly and become unusable. By contrast, these colours could be reused by just adding water and are very economical to make.

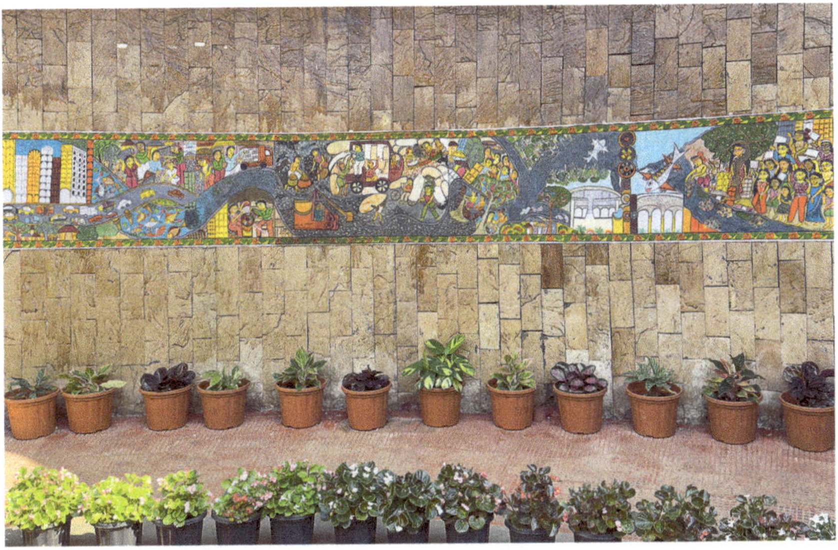

Fig. 3.3 Mamoni Chitrakar, The Mithi River Scroll, 2023. Maharashtra Nature Park, Dharavi, Mumbai. (Image: Ben Parry)

The story of Mamoni's own life as an artist and the production of the scroll, and how it enabled a big debate about the histories of Dharavi, the Mithi river, and the past, present, and future of that space exemplifies how the residency programme at the lab moved beyond a conventional art curriculum. It wrapped its audiences in the traditional craft of oral storytelling as a profound piece of work rooted in performance and exchange—whose approach to art making and fostering understanding meant that those who saw it were immediately able to relate to it. The fact that Mamoni was well travelled (the young people with smartphones had quickly researched her accomplishments) but spoke and sang mostly in Bengali made her 'one of them', and she became an instant role model.

In April 2023, as part of showcasing the work of the Lab, the scroll was exhibited at the Maharashtra Nature Park, alongside a video of Mamoni performing the scroll, enabling public dialogue with visitors and the many school groups in attendance. A green oasis providing education and awareness about nature conservation, the Maharashtra Nature Park runs along the banks of the Mithi River on the northern edge of Dharavi. Reclaimed from a former waste dumping ground, the park speaks of the histories of waste and the potential for ecological restoration. The scroll provided narratives in which the river and the nature park were part of learning. The exhibition raised many questions about the role of the park: what is nature, what is not, what does it mean to live next to a polluted river? How do the ecologies of the city and local ways of life connect to issues of biodiversity? All our projects set out to connect with nearby resources, bringing artefacts, objects, stories, and people into a space to generate new learning. We describe this as *building shared knowledge platforms* using arts-based processes in which everyone can observe and participate. Mamoni's approach—singing to the scroll—also creates a certain sense of ritual, distance, and objectivity, which the young people can relate to, but which is also 'other'. These insider/outsider relationships are important because they give permission to think differently, more boldly, and perhaps more openly and interculturally.

Lab as 'Urban Assemblage'

People who live in Dharavi have a strong sense of identity: there is always an awareness of 'we' and 'us'—of inside/outside relationships, and it is always clear who is from outside. This strong sense that 'Dharavi is ours' is built on decades of intergenerational struggle to carve out viable spaces to build livelihoods on five acres of occupied, reclaimed land. The sheer density and intensity of social, spatial and economic relations generates many different forms of solidarity and self-help.

Alongside ACORN, the Lab opens alternative possibilities and pathways to break the cycle of being drawn into exploitative labour and present alternative options for earning and learning. Simply left to individual families, economic imperatives usually mean that by the age of 13 or 14 young people either must join a family business or get married. ACORN and the Lab's work creates opportunities to learn new skills and widen choices. Within Dharavi there are opportunities to connect and learn: through schools, projects, NGOs, and industries, because of the geographical advantage of its location and connectivity. This homegrown ecology of learning and human resources is likely to be lost if the Dharavi Redevelopment Plan goes ahead in its current form.

These forms of connective pedagogy occupy a pivotal role in the realm of contemporary public art, community-based art, and situated practices. The Lab has served as the foundation for collaboration and co-production of knowledge and the creation of a safe space that facilitates the mediation of societal differences. Pedagogy in this context involves a dynamic process where both teacher and learner engage in simultaneous learning and unlearning. They collaborate in the pursuit of knowledge production, forming communities through ongoing interactions, and the shared utilisation of space, expertise, and labour, all aimed at fostering much-needed socio-political, sustainable environmental change. Within this framework, the artist assumes various roles, serving as an organiser, creator, catalyst, or empathetic collaborator. This environment provides secure spaces for mediating caste, class, and gender differences through processes of empathy, care, and repair.

In these ways, within the work of Compound 13 Lab, notions of life, livelihood, learning, work and art all come together. It is also a place, under the umbrella of ACORN's work in general, that young people want to come to and 'hang out' in, much in the tradition of youth centres. Young people who attend Compound 13 Lab report that this atmosphere of acceptance and inclusion contrasts with their experiences of formal schooling. Most of what is taught in schools seems very distant from the kinds of lives that they are living, whereas the Lab directly addresses their circumstances. Our notion of a living curriculum, built upon the principle of designing learning frameworks which are not fixed, but open, adaptable and dialogical, was an essential part of imagining how the Lab might function: where informal urban knowledges, rooted in the everyday experiences of people living in neighbourhoods like Dharavi, intersect with other opportunities for shared learning and research. We consider the Lab as a physical place of possibility, a space for thinking and making, underpinned by a set of relationships that offer possibilities for imagining, living differently and doing things differently. Deployed here art creates a space of dialogue, shared learning and knowledge making, enabling us to work together through different ways of knowing, across languages and cultures, and to develop a sense of creative agency.

Conflict of Interests (1) All authors declare that they have no conflicts of interest. The work underpinning this chapter was funded by the following research grants:

AHRC: AH/P00637X/1 Resources of Hope: Giving voice to underprivileged communities in India (2016–2018, AHRC/GCRF)

AHRC: AH/S0035897/1 Rethinking Waste and the logics of disposability: Compound 13 Lab (2019–2020, AHRC/GCRF)

British Academy: UWB 190022 Waste, Water and Wellbeing: lessons from the interface of formal/informal urban systems in Dharavi (2019–2023).

References

Banaji, S. (2024). Against resilience: The (anti-)ethics of participation in an unjust and unequal public sphere. *Javnost – The Public, 31*(1), 141–157.

Boyle, D. (2008). *Slumdog millionaire*. Film4/Celador Films.

Chhabria, S. (2019). *Making the modern slum: The power of capital in colonial Bombay*. University of Washington Press.

Doron, A., & Jeffrey, R. (2018). *Waste of a nation: Garbage and growth in India*. Harvard University Press.

Gidwani, V. (2015). The work of waste: Inside India's infra-economy. *Transactions of the Institute of British Geographers, 40*(5), 575–595.

Gill, K. (2010). *Of poverty and plastic: Scavenging and scrap-trading entrepreneurs in India's urban informal economy*. Oxford University Press.

Jeffery, G. (2005). *The creative college: Building a successful learning culture in the arts*. Trentham Books/IoE Press.

Jeffery, G., & Parry, B. (Eds.). (2023). *Waste work: The art of survival in Dharavi*. Wunderkammer Press/Art Editions North.

Joshi, P., Matthew, I., & Mehrotra, R. (2011). *Mumbai reader 10*. Urban Design Research Institute.

Kelly, H., & Jeffery, G. (2016). Community media and cultural politics on Tyneside. In P.'s. K. E. Collective (Ed.), *People's knowledge and participatory action research* (pp. 73–82). Practical Action Publishing.

Leeson, L. (2017). *Art: Process: Change: Inside a socially situated practice*. Routledge.

McCloud, K. (2010). *Slumming it*. Talkback Thames/Channel 4 Television.

National Geographic. (2007). Dharavi: India's shadow city. *National Geographic, 211*, 5.

O'Hare, P. (2022). *Rubbish belongs to the poor: Hygienic enclosure and waste commons*. Pluto Press.

Pancholi, V. S., & Kamble, L. (2023). *Dharavi waste flows map: Mumbai*. Compound 13 Lab.

Parry, B. (2015). *Cultural hijack: Critical perspectives on urban art intervention*. PhD. University of the West of Scotland.

UN Live, Museum for United Nations. (n.d.). Retrieved August 7, 2024, from https://www.museumfortheunitednations.com

Wakeford, T., & Sanchez Rodriquez, J. (2018). Participatory action research: Towards a more fruitful knowledge. In K. Facer & K. Dunleavy (Eds.), *Connected communities foundation series*. University of Bristol/AHRC Connected Communities Programme. Retrieved August 13, 2024, from https://ore.exeter.ac.uk/repository/bitstream/handle/10871/36994/Action_Research_SP.pdf?sequence=1&isAllowed=y

CHAPTER 4

'Dokkhina Sundari': A Local Story of Climate Emergency in Bangladesh

Aminul Hoque and Sudip Chakroborthy

Introduction

This chapter outlines an in-depth conversation in May 2024 between Dr Aminul Hoque, a Bangladeshi-born academic and ethnographic researcher who now lives in East London, UK, and Dr Sudip Chakroborthy, an arts practitioner from the University of Dhaka, Bangladesh. The key focus of this conversation is centred around the 2014 production of 'Dokkhina Sundari', a play that was written by Shahman Moishan, directed by Sudip Chakroborthy and produced by TheatrEX Bangladesh. It was created with the National Theatre of Scotland for The Tin Forest Theatre Festival which was part of the

A. Hoque (✉)
Educational Studies, Goldsmiths, University of London, London, UK
e-mail: a.hoque@gold.ac.uk

S. Chakroborthy
Theatre and Performance Studies, University of Dhaka, Dhaka, Bangladesh
e-mail: sudip@du.ac.bd

M. Matthews (ed.), *Ecologies in Practice and Learning: Arts Interventions in the Earth Crisis*, Palgrave Studies in Educational Futures, https://doi.org/10.1007/978-3-031-82338-1_4

Glasgow 2014 Commonwealth Games Cultural Programme, and was supported by the British Council. The production toured Bangladesh, India, England and Scotland in 2014.

'Dokkhina Sundari' translates to 'southern beauty' and refers to the endangered Bengal Tiger, often found in the Sundarbans region of southern Bangladesh. The Sundarbans, which is a UNESCO world heritage site, is also noted for its unique bio-geo-cultural diversity, is considered a 'green wall' for the coastal region, and also has the world's largest mangrove forest. The protection of this forest is essential as it acts as a natural defence and buffer zone against hurricanes and floods. 'Dokkhina Sundari' is a story of defiance and resistance, shares stories of the Sundarbans and raises a critical voice against the Bangladeshi government's decision to build the coal-based thermal Rampal Power Station very close to the Sundarbans. This project has also received much wider condemnation from the international community and local pressure groups for its environmental impact and 'serious threat' to the Sundarbans (UNESCO, 2016), especially on local wildlife, habitat and the many millions of local indigenous people who depend on the Sundarbans region for their livelihoods, such as woodcutters, fishermen and honey cultivators.

As a theatrical performance, 'Dokkhina Sundari' is a colourful and creative expression and parable of this dissent against deforestation and the crisis of corporatised globalisation. Aimed at a local audience, national and international artistic community, and anyone who is interested in wider issues of social and environmental justice, the performance is a myriad of creative expressions—incorporating narrations, local oral histories and myths, folk and fairy tales, conversations, indigenous and contemporary choreography and dance, prose, poetry and music. We find out more about the everyday lived experiences of the people who are impacted by this environmental crisis, especially those who are impoverished, isolated and vulnerable. The ultimate message of 'Dokkhina Sundari' is a simple, yet important one—that *all* life is precious (humans, animals, plants) and that we *all* have a moral and ethical duty to preserve this.

The story of 'Dokkhina Sundari' is significant as it highlights the grave importance of climate emergency and climate justice in Bangladesh. The production also illustrates the overlap between the artistic community and the environment, and brings attention to the critical role that the arts can play in raising awareness of wider issues of social justice. Importantly, productions such as 'Dokkhina Sundari' provide a medium to communicate a wider message to a more global audience, and in the process, there is a social justice imperative in that it gives humanity and voice to groups of people, plants and animals, who are often voiceless, powerless and invisible.

The Climate Emergency in Bangladesh—Wider Context

In recent times, the climate emergency in Bangladesh has received much attention from international organisations (United Nations, 2024) and has also attracted wider media and scholarly focus (Channel 4 news, 2016; The Economist, 2022).

Identified for Official Development Assistance (ODA) by the United Nations (2024), Bangladesh is one of the world's poorest and most densely populated countries; home to over 160 million people in a country just over twice the size of Ireland. Due to its topography, geographical location and rapid urbanisation, Bangladesh remains at the forefront of all concerns and discourse related to climate change. Global warming causes climate change, contributing to changes in (extreme) weather patterns and seasons around the world. Global warming has also contributed to the rise in sea levels caused by the expansion of warmer seas and melting ice sheets and glaciers. This poses a serious threat to life on Earth in the forms of widespread flooding and extreme weather (National Geographic, 2024a). Bangladesh is facing multiple climate change impacts, including rising sea levels, widespread poverty, hunger and malnutrition, heat stress, access to clean water and drought and shattering natural disasters often causing displacement and homelessness on a large scale (Chowdhury et al., 2022).

Bangladesh is low-lying and vulnerable to persistent flooding. It stands to be badly affected by any rise in sea levels. Three of the world's largest rivers (Ganges, Brahmaputra, Meghna) join within Bangladesh, creating one of the world's largest floodplains. The average temperature in Bangladesh is predicted to rise from 1.0 °C to 1.5 °C by 2050 (Mani et al., 2018) and some scientists estimate that by 2050, 17% of the country will be submerged by rising sea levels leaving 20 million people without a home (National Geographic, 2024b). Furthermore, the country is subject to numerous hazards including devastating cyclones: it is ranked first in the world in terms of vulnerability to tropical cyclones (see Mim, 2021). In addition, although women have a significant role to play in helping communities cope with the effects of climate change, cultural customs and practices often work against this. Therefore, recognising women's active contribution and supporting gender equality as a critical strategy in global adaptation to climate change is also a key challenge (Alston, 2015). Amidst all of this, Bangladesh is hosting approximately one million Rohingya people who fled violence in Myanmar, in what has become the world's largest refugee camp.

In response to this climate emergency, the Bangladesh government has formed a number of initiatives to counter the climate crisis outlined in the Mujib Climate Prosperity Plan (2021) focused around innovative technology, closer collaboration with international agencies and countries such as the United Kingdom, adaptation to changing labour markets, population control through better health and education, a concentration on the vulnerable members of society and renewable 'green' energy that is climate-resilient, low-carbon, gender-responsive and socially inclusive. Others have noted that despite many geo-political challenges, Bangladeshis have 'trailblazed' (Davison, 2022), shown 'resilience' (Alom, 2022) and 'adapted' (Ayers et al., 2014) when it comes to fighting disasters through (predominantly female) volunteering networks, developing early warning systems for cyclones and storms and making better use of satellite technology, community-led action, building more cyclone shelters and homes on stilts with stronger long-lasting materials to withstand extreme weather, floods and mudslides, new and innovative farming and fishing techniques, reinforcing eroding river banks and so on.

Within this wider geo-political context, 'Dokkhina Sundari' offers a critique and also raises local, national and international awareness to the climate emergency in Bangladesh. It has been suggested that while the country is keenly aware of its vulnerability to climate change, not enough has been done to match the pace and scale of the resultant displacement and urbanisation (McDonnell, 2019). In the following conversation, that took place exactly 10 years from its first showing, this interview asks some hard-hitting questions of the aims and objectives of this production, the key message, who the audience was/is, and also the wider legacy of the play (Fig. 4.1).

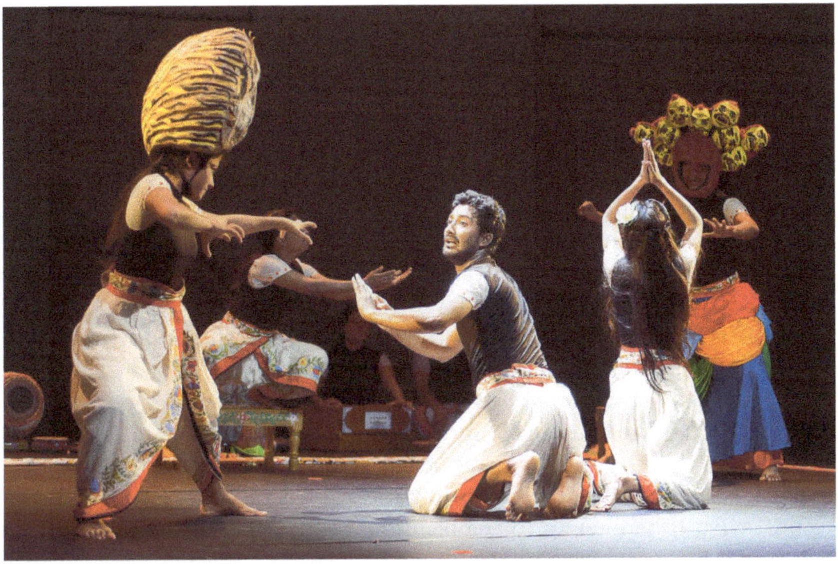

Fig. 4.1 Dokkhina Sundari 2014. (Photograph Tim Morozzo)

The Conversation

Aminul: Before we start, tell me a little bit about yourself and your background? How did you become an artist and an academic?

Sudip: Theatre, music, artistic performance has been in my family and it is something that I was exposed to from a very young age as both of my parents are artists. I was interested in these [pala] traditional [Bangla] performances, music and songs since my childhood. When I started my study at college, I became involved with a local Theatre Group named *Theatre Shakha-Borak* in Nabiganj in the district of Habiganj. Later, I got the opportunity to study theatre at the University of Dhaka, Bangladesh. I started my teaching in the same university as soon I finished my BA and MA studies. Subsequently, I became a theatre practitioner in the public domain.

Aminul: It is really useful to know more about your background story. Let us jump straight into the 2014 production of 'Dokkhina Sundari'. Tell me more about the story behind this production and also what the story symbolises? What is the literal translation of 'Dokkhina Sundari'? What does it mean to our non-Bengali speakers?

Sudip: 'Dokkhina Sundari' means *the beauty of the South* of Bangladesh. The play theatricalises the epic canvas of primordial living and eternal aspirations in southern Bangladesh which is famous for the Sundarbans—a region which has the world's largest mangrove forest and is noted for its unique bio-geo-cultural diversity.

In response to the government's decision to build a coal-based power plant in the Sundarbans territory—a 'green wall' of the coastal area of Bangladesh protecting the country from tornadoes and cyclones coming inland from the Bay of Bengal—I took the plan to stand against this step. So the story highlights the struggle of natural beauty against man-made threat, and also raises awareness related to the endangerment of the Royal Bengal Tiger. The play talks about the danger of deforestation to the Sundarbans and the local environment and communities, habitat and animals. I see this theatre production as a way to voice a form of dissent against what is happening to the beautiful South. The story of 'Dokkhina Sundari' is not just for a national audience, neither is it solely a message for government. The importance of the climate emergency in Bangladesh is also of importance for a wider international audience. … When we got the opportunity to tell our story to an international audience during the Commonwealth Games of 2014 in Scotland, we decided that this was the best time to tell our story, because this play is about us, our country, and also about our future. … In this production, I had the opportunity to work with many academics, and a playwright who contributed to the drafting and writing of the story.

Aminul: You are from Sylhet—the Northeast part of Bangladesh, but you created 'Dokkhina Sundari' that is based on Sundarbans—the Southwest part of the country. What is your connection to that region? I know there is a national connection, but geographically there is quite a big disparity. Why not tell a more localised story based on where you are from?

Sudip: To me as an artist, it is important to tell the stories for any and every part of the world. I think, the story of Sundarbans is now a very important one not only for Bangladesh, but also for the world. See, we are now facing record temperatures in Bangladesh. The winter season is gradually shortening its length. Right now, I am discussing with you over a Microsoft Teams call from Bangladesh and right now the temperature of the country is more than 40 degrees Celsius. But just last year, the average temperature was 38 degrees Celsius at this time, and five years prior, it was

35 and just 10 years ago, it was only around 30–32. So the temperature is increasing every year, and we are going through many different known, unknown and unidentified environmental issues. It has impacted our welfare, health, economics, livelihoods, human behaviour. ... Clearly, it is now a global issue. No doubt, we are in a serious global climate crisis. So, from an artist's point of view, it is not connected with my place of origin, but connected with wider concerns. That is the reason, I preferred to talk about the issues in Sundarbans when the power plant project started. At that time, I was thinking, it is the right time to raise the artist's voice to the local and the global.

Aminul: Take me back to the actual production. As in what were your key memories of the production and the performance? What are your fondest memories?

Sudip: That I was able to tell a local story about a global issue. And this local story gave voice to local people, and was a fusion of local myths, fairy tales, parables, folk and traditional music and art, contemporary dance. ... And that this performance allowed our voices and dissent against corporatised globalisation which is impacting so many people around the world.

Aminul: I can see that this is a very personal story for you, because as you mentioned a few moments earlier, you're in Bangladesh right now, it is early May and you're saying the temperature's almost 40. Alongside the rising temperatures, geographically, Bangladesh is also a region of the world that is prone to lots of cyclones and typhoons. There is immense poverty around the country, mainly in the rural areas. So, it means that lots of people are displaced in the rural farmlands and they subsequently move into the capital (Dhaka), and other major cities, and the cities become more overcrowded. And then there are more issues of plastic pollution, overcrowding, homelessness, crime and poverty. So this is a vicious cycle that the climate emergency is contributing towards. You are very much viewing this as a story, not just about Bangladesh, but also a personal story because you are experiencing it and not just you, but many of your family and communities are also experiencing this and living through the climate emergency right now.

Sudip: Yes, yeah.

Aminul: Was the production first shown in Glasgow or was it shown in Bangladesh beforehand?

Sudip: It was premiered at the University of Dhaka in June 2014, then toured the UK and India. In the UK, it was showcased for the East Londoners first, then in an international theatre festival in Manchester.

Finally, 'Dokkhina Sundari' showcased at the Tin Forest Theatre Festival which was part the 20th Commonwealth Games 2014 in Glasgow. … Bangladesh heritage community members attended the shows in London, Manchester, and Glasgow.

Aminul: When your production was viewed by a national and international audience, what was the reception like? What was the feedback? What was the commentary? What did people think of it?

Sudip: At the end of the show, we got a standing ovation! We think, people liked it, they liked our performance. When we met people, after the show, they said, it is a timely production and well-articulated. The story travelled towards the essence of pantheism. In response to the authority's decision to build a power plant in the Sundarbans territory, the performance was considered *timely*. People liked the content of the play and the presentation style that includes traditional Bangla music, dance, movement, and acting.

Aminul: I did not see it live during the time of 2014, but I have seen the subsequent video that you have sent me, and I must say it is very immersive. It is very creative. It tells a really important story through the media of dance, music, and poetry, like you mentioned. I wonder, and this is this is not a critique, but it is more intrigue from my end. I wonder who you thought would be the main audience? Of course there is a national and international audience but break it down for me a bit more. Is it just the artist community and the student intellectual community that you envisaged seeing your production and engaging with it, or was there a wider mass audience involved as well?

Sudip: In Bangladesh and India, mostly the audience were students, teachers, academics, artists and environmental activists. In the UK shows, we got, Bangladesh heritage community members as well as members from the international artist community.

Aminul: I wonder whether you had the scope at that time—because you said it was a *timely* production—to have taken the production on a road journey to the more rural parts of Bangladesh, too hard to reach, disconnected and alienated parts of the country? We know Bangladesh is mainly rural and agricultural and these are the very people that are affected and impacted most by the climate change.

Sudip: Unfortunately, I was not able to do this because of time frame, money, and yeah local support. So, I only did it in the theatres, not in the open spaces. But we have the scope to connect with a wider audience via the digital version of the play.

Aminul: Yeah, we know now that the world is becoming much more connected through digital media. There are many isolated parts of the world who do access these very important messages through technology. So hopefully your message via your production is now seen by a mass wider audience. The journey of an artist has the power to communicate his or her ideas and message to a much wider audience.

Sudip: Yes. I agree.

Aminul: But unfortunately, budget and infrastructure are two of the key obstacles that get in the way.

Sudip: Yes, exactly! ... The country is keenly aware of its vulnerability to climate change but not enough has been done to match the pace and scale of this change. ... To tackle this situation [of displacement], we have very limited initiatives. What we have is urbanisation, rapid infrastructure and development of new buildings. Despite these kinds of globalised, new liberalised, corporatised initiatives, we have very limited initiatives to make the country greener.

Aminul: Yes, and as part of this resistance, like you mentioned earlier, you are raising your voice of dissent and consent through the medium of performance and theatre, which you are very much involved in from the Bangladeshi perspective. Of course, you are not claiming that your voice and your performance are changing the narrative; but you are definitely contributing to a wider discussion and a discourse, a critical discourse towards this climate emergency that we are seeing, not just in Bangladesh but across the world. I think you and I are both very much aware that whatever intentions you have, they are limited in scope because of the wider challenges involved. But it is clear that you have not given up and you continue to pursue this message through all of your social justice work, in academia, in theatre, in performance, at community, at local, national and international level.

Sudip: Yes, indeed.

Aminul: I wanted to go back to a question I meant to ask earlier: did you, in the course of the production, manage to go and do some field work in the Sundarbans? Did you actually go and visit, meet with locals, speak to them? Did their stories and lived experience inform the actual writing and the production of the performance?

Sudip: It was the starting point of our writing, rewriting, making, and creating 'Dokkhina Sundari'. We often visited the southern part of the country. It is to be also noted that I had similar experiences [of flooding, rising temperatures] in my childhood. ... So, in fact, the play is created from my own lived experience. This was also further informed through my own ethnographic visits to the Sundarbans, living there, meeting local people and actively engaging with local communities. ... It was important that we got to hear *their* stories from *their* perspective.

Aminul: Okay [as a sociologist and ethnographer myself] I'm very heartened by this. The knowledge that your play is not just informed by your own lived experience, but also the ethnographic immersion into localised lived experiences and spending time with the very people that you are writing about. ... Now, I want to go further into the world of artists like you, especially living in a country like Bangladesh. In your own words, what role can the artist community play in tackling and or raising awareness of this climate and social justice emergency that we are experiencing in Bangladesh? What do you think should be the role of artists?

Sudip: Artists are not silent but sometimes they are silenced. Amid wider rapid urbanisation and globalised corporatisation, we all have a duty to raise a critical voice and think about the environment and climate.

Aminul: Having watched your production, it is a theatrical critique of wider government environmental policy, and it also examines the critical role of big business and the interplay between business, policy and politics. So, with that in mind, what has been the reaction from business and government to your actual play? Did you receive any feedback? Did you receive any criticism? Were there some obstacles put in front of you, or can the artist community in Bangladesh critique, comment on, disagree with wider initiatives, especially in relation to government?

Sudip: From my own experience, the Bangladeshi government do not care much about what the artist community thinks. The theatre and performing arts community in Bangladesh is limited in audience, especially for indigenous and traditional theatre. They do take seriously any points of view from the artist community. [Mainstream creative industries] such as film, media, print press are taken far more seriously. Theatre buildings only have 300 to 400 audience capacity, most of which are from the intellectual community. ... 'Dokkhina Sundari' had many newspaper reviews, in-person feedback, and we engaged in many live after show Q&A sessions with the audience. One review that must be mentioned was written by Abdul Gaffar Chaudhury who wrote a sub editorial for a national English newspaper in Bangladesh. From policymakers in Bangladesh, we did not receive any comments.

Aminul: To be honest, the critique of government, the critique of policy, the critique of big business, the critique of big capital, through discourse, through academia, through journalism, through theatre and performing arts, through literature, through critical voice, and now through social media. ... This is not just limited to Bangladesh, this is a global situation, right? In the UK context, we often use social and digital media, the creative industries, journalism, academia—which I'm very much a part of—we often use these media as platforms to voice dissent against big institutions and power structures. So, what you are doing is no different to what is happening globally, right?

Sudip: Yes, these are global issues.

Aminul: It has been 10 years since the production, I just want to move on and ask, what do you think are the wider social, political, and educational messages of productions such as 'Dokkhina Sundari'?

Sudip: There is a profound social message in that everyone has the right to think and speak. 'Dokkhina Sundari' not only allowed us to put our point of view forward, from the artists' point of view, but it also allowed a wider community message of climate social justice to be put forward. The community has the right to a voice. So, I think, our initiative created a space which invited and allowed people to discuss, debate and think. I was pleased to also see so many young people as part of the audience, as this is about their future also.

Aminul: One of your key objectives behind 'Dokkhina Sundari', chimes in with a key reason as to why I am in academia—not to have answers or solutions but to make people *think* and to develop the confidence to *raise their voice and ask questions*, and engage in debate and discussions about important issues that do not only impact them but also impact others locally and internationally.

Sudip: Yes. I completely agree with you.

Aminul: I want to know more about the Glasgow Commonwealth Games of 2014. How was the feeling when you found out that the production was going to be part of this event?

Sudip: Yes, I was excited. It was an exciting adventure for me and for Bangladeshi theatre, since it was the first ever theatre production invited for the Commonwealth Games Theatre Festival. So, it is an important and memorable moment for me as a theatre maker, as an artist.

Aminul: How was the experience of Glasgow 2014—being part of this international event?

Sudip: When we got the standing ovation after the show, we thought, we did it! I was amazed, and it made me cry that we made it after a long struggle with funding and visa issues. We had some other shows in London and Manchester, which was not part of the Tin Forest Theatre Festival, therefore we also had logistical issues around travel and accommodation. [Showing the play to the UK diaspora Bangladeshi community was also important] because I wanted to raise awareness of these issues with people who have Bangladeshi heritage so that they can get more connected with their *home*.

Aminul: So, what you're saying to me is that it was important for you to take this message of what is happening in the motherland to the Bangladeshi diaspora because we need to be aware? I'm part of this diaspora and we need to be also more connected to issues happening many thousands of miles away.

Sudip: Yes.

Aminul: The final question, what do you think is the legacy of 'Dokkhina Sundari'?

Sudip: The first legacy is people still remember this production. People liked it. They think about the context and interpretation we performed in 'Dokkhina Sundari'. In terms of the context of the play, that legacy is, I think the question and the discussion we wanted to create. We created it and we did it.

Aminul: What is the question that you wanted to create?

Sudip: The question is: is it necessary to build this power plant in the Sundarbans, to the detriment of the local people, animals, environment and habitat, especially when we have a climate emergency? In 2014, the same year of our theatre production, we saw, many protests and marches, many political and artistic discussions and debates that took place across Bangladesh. … Ultimately, in the play 'Dokkhina Sundari', we wanted to echo the philosophy that life is important. This includes *all* life—humans, nature, animals, birds, trees, insects etc. Life is important. We need to maintain the biodiversity and protect our environment. If we destroy it, we destroy ourselves (Fig. 4.2).

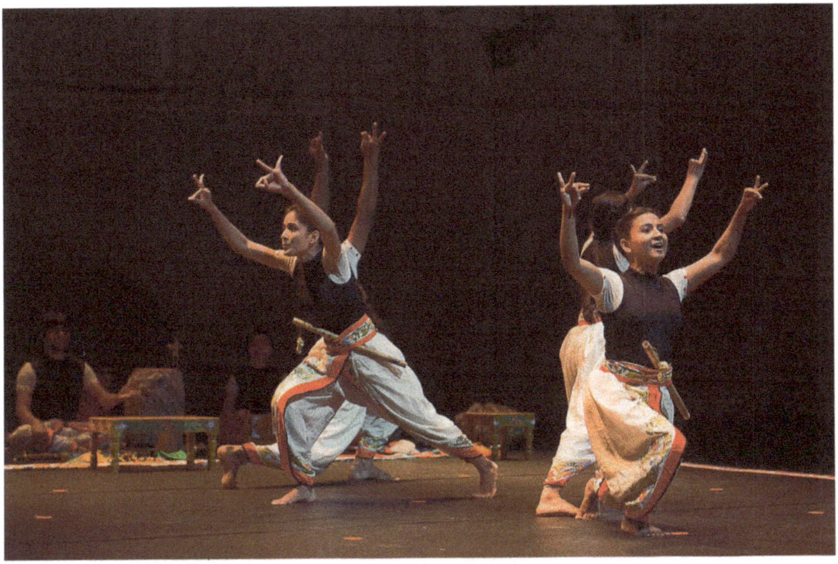

Fig. 4.2 Dokkhina Sundari 2014. (Photograph Tim Morozzo)

Conclusion

It feels appropriate to conclude this chapter with this key message from Sudip towards the end of the conversation, that 'all' life—humans, nature, animals, birds, trees, insects—is important, and 'if we destroy it, we destroy ourselves'. The story of 'Dokkhina Sundari' is both educational and significant as it underlines the grave importance of climate emergency and climate justice in Bangladesh. The production also highlights the complex intersection between the creative arts and the environment, and reiterates the critical role that the arts can play in raising awareness of wider issues of environmental social justice. Crucially, productions such as 'Dokkhina Sundari' provide a platform to communicate a local story to a more global audience, across distance, language and cultures, and in the process, we learn from each other about universal issues that affect us all. Therefore, there is also a social justice imperative in that it gives humanity, visibility and voice to groups of isolated, alienated and disadvantaged people, plants and animals, who are often voiceless, powerless and invisible (Fig. 4.3).

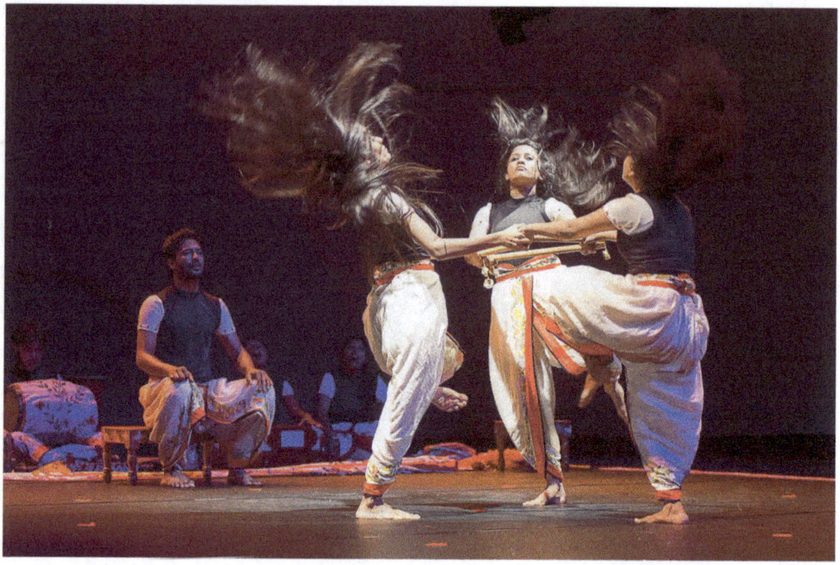

Fig. 4.3 Dokkhina Sundari 2014. (Photograph Tim Morozzo)

Conflict of Interests (4) The authors have no conflicts of interest to declare that are relevant to the content of this chapter.

References

Alom, Q. (2022, June 1). The people of Bangladesh are resilient, but without urgent climate action, the future of this fascinating country is uncertain. *International Fund for Agricultural Development*. Retrieved May 15, 2024, from https://www.ifad.org/nl/web/latest/-/aunque-los-habitantes-de-bangladesh-sean-resilientes-de-no-tomarse-medidas-urgentes-contra-el-cambio-climatico-el-futuro-de-este-fascinante-pais-sera-incierto

Alston, M. (2015). *Women and climate change in Bangladesh*. Routledge.

Ayers, J. M., Huq, S., Wright, H., Faisal, A. M., & Syed, T. H. (2014). Mainstreaming climate change adaptation into development: A case study of Bangladesh. *WIREs Climate Change, 2014*(5), 37–51.

Channel 4 News. (2016). *Climate change: How can poor countries grow without pollution?* Retrieved June 6, 2024, from https://www.youtube.com/watch?v=-h6_hJFD7to&t=274s

Chowdhury, M. A., Hasan, M. K., & Islam, S. L. U. (2022). Climate change adaptation in Bangladesh: Current practices, challenges and the way forward. *The Journal of Climate Change and Health, 6.* https://doi.org/10.1016/j.joclim.2021.100108

Davison, C. (2022). The country trailblazing the fight against disasters. *BBC.* Retrieved June 6, 2024, from https://www.bbc.com/future/article/20220719-how-bangladesh-system-fights-cyclones-climate-disasters

Mani, M., Bandyopadhyay, S., Chonabayashi, S., Markandya, A., & Mosier, T. (2018). *South Asia's hotspots: Impacts of temperature and precipitation changes on living standards.* South Asia Development Matters: World Bank Group.

McDonnell, T. (2019, January 24). Climate change creates a new migration crisis for Bangladesh. *The National Geographic.* Retrieved June 9, 2024, from https://www.nationalgeographic.com/environment/article/climate-change-drives-migration-crisis-in-bangladesh-from-dhaka-sundabans#:~:text=Interviews%20with%20dozens%20of%20migrant,of%20the%20resultant%20displacement%20and

Mim, M. M. (2021, May 24). The concise history of the 10 deadliest cyclones in Bangladesh. *The Business Standard.* https://www.tbsnews.net/environment/concise-history-10-deadliest-cyclones-bangladesh-250063

Mujib Climate Prosperity Plan. (2021). *Mujib climate prosperity plan: Decade 2030.* https://mujibplan.com/wp-content/uploads/2021/12/Mujib-Climate-Prosperity-Plan_ao-21Dec2021_small.pdf

National Geographic. (2024a). *Global warming.* Retrieved March 3, 2024, from https://education.nationalgeographic.org/resource/global-warming/

National Geographic. (2024b). *Environmental refugee.* Retrieved June 6, 2024, from https://education.nationalgeographic.org/resource/environmental-refugee/

The Economist. (2022). *See what three degrees of global warming looks like.* Retrieved June 6, 2024, from https://www.youtube.com/watch?v=uynhvHZUOOo

UNESCO. (2016). *World Heritage Centre and IUCN call for relocation of Rampal power plant, a serious threat to the Sundarbans.* Retrieved June 7, 2024, from https://whc.unesco.org/en/news/1573

United Nations. (2024). *The United Nations in Bangladesh.* Retrieved March 3, 2024, from https://bangladesh.un.org/en/about/about-the-un

PART II

Practice Research and Arts Activism in Universities

This part of the book engages with ecological practice research with participating learning groups, in or connected with Higher Education Institutions (HEIs). Authors foreground explorations through arts practice, and research narratives that involve making, doing, exploring, imagining and inventing. They argue that ecological practice research helps materialise alternative possibilities to standardising neoliberal and Earth occluding forms of education. Researchers in HEIs are reaching out to communities of young people and teachers, also looking into their own collaborative ecological arts practices.

Miranda Matthews questions how arts practice can help address eco-anxieties and empower young people, both in learning spaces and in complex societies. Matthews discusses how arts methods empower young people to voice eco-anxieties with focused and liberated expression, and to learn different sensory approaches to making interventions in the Earth Crisis. The Ecologies in Practice project developed by the Centre for Arts and Learning and Climate Museum UK has connected practice researchers with schools, young people, teachers and university students. Matthews raises issues for becoming empowering arts educators without being didactic, and brings climate awareness into a synthesis of posthumanist, indirect and critical pedagogies.

Francis Gilbert and Anna Stewart show how young people can form critical ecological agency, in presenting the Parklife project. Parklife forms collaborations among young people, with university students, teachers and academics, to bring about positive changes in urban environments.

Gilbert and Stewart discuss how arts practice can activate the ecologically aware connections that young people have with the green spaces of parks. They work with creative research methodologies that enable young people to mediate a tragedy, and to reclaim creative and safe connections with nature in 'learning walks' that deterritorialise the park as an assemblage of interactions between human and other than human life forms.

Emese Hall documents a large scale collaborative international project: CARE: Connecting Art with REal Life Issues. Hall gives an account of how this empathic, ecological project has engaged with the climate related Global Sustainable Development Goals of the United Nations Educational Scientific and Cultural Organisation (UNESCO). Exploring how the arts can help protect nature, and develop ecologies of practice among primary school teachers, Hall positions a spatialised, multinational and architectural educational practice, in a discursive relation to sociological ecosystems. Hall analyses usage of the term 'ecology' in Europe, and connects with the visual expressions of the need for ecological action that are represented in children's artworks.

Ben McDonnell and Molly Pardoe offer an interpretive and subtle view of ecologies in practice research, that is placed in an 'in-between' space of doctoral studentship within the hierarchies of a university. New-materialist, affective approaches challenge the neoliberal performativity of the establishment, with alternative values for the collaborative and playful 'marginal' spaces of resistance enabled by arts practice research. The authors view happenings in their making spaces as 'noticings' in an equitable creation of 'the ecology of the in-between'. Bringing hopeful pedagogies into practice, McDonnell, Pardoe and Dixon question and re-position a view of how sustainability relates to play and collaboration.

These projects are viewed to be taking creative action, or engaging participants in arts focused approaches to activism in the Earth Crisis. The projects represented connect with groups of young people, and offer alternative posthuman practices that centralise youth, student and practitioner voices. Practice research among artist educators who explore boundaries and borders in ecological making processes, and ways of activating aesthetic imaginaries, is also included at this stage.

CHAPTER 5

Arts Methods in the Earth Crisis: Empowering and Learning from Youth Voice

Miranda Matthews

Introduction

In most regions of the world young people know that there is an Earth Crisis (Hickman et al., 2021). Millions of students have been on a school strike to convince adults in power to take urgent action to redress climate injustice (Boulianne et al., 2020). Eco-anxiety is known to be affecting young people and adults (Verlie, 2021), but there was a gap in research of how arts practice could enable young people to express themselves in these emotive global conditions. The Centre for Arts and Learning and Climate Museum UK developed practice research to investigate how arts methods could empower youth voice in connection with schools.

In this chapter I will discuss research that explores how artists and arts educators can centre creative practices, to counteract the man-made causes of the crisis, and take humans and other-than-humans beyond the worst of the Anthropocene; this is the era defined by the impact of humanity on the

M. Matthews (✉)
Educational Studies, Goldsmiths, University of London, London, UK
e-mail: M.Matthews@gold.ac.uk

M. Matthews (ed.), *Ecologies in Practice and Learning: Arts Interventions in the Earth Crisis*, Palgrave Studies in Educational Futures, https://doi.org/10.1007/978-3-031-82338-1_5

planet (jagodzinski, 2018). In 2021–2022 I curated an Ecologies in Practice research programme, and invited Bridget McKenzie from Climate Museum UK to speak. We therefore had the beginnings of a fruitful research partnership in the making. With support from the Arts and Humanities Research Council, and Goldsmiths University of London, CAL and CMUK practice researchers created an evolving methodology that amplifies the voices of young people.

Participatory Arts Methods for Engaging Young People in Climate Research aimed to consult young people in discussion and making. We wanted to research how interventions in arts practice could increase their agency around climate change, extractivism, biodiversity loss and cultural relationships to the environment. The research team wanted to create dialogical interventions that would connect with the needs of different learning communities. We planned to resource participants with three arts methods for engaging with Earth Crisis information. Our workshops would happen outside the school environment in a university space to potentially liberate playful, social creative voice. The workshops were to activate freedom of self-expression, encourage respect for difference and an ethics of care for interspecies connections (Atkinson, 2018).

Climate Museum UK's Listening Project (2021) had sounded out the feelings of young people and their demands for adults to makes changes. In addition, Francis Gilbert and I had written about how affective free writing and drawing (Fig. 5.1) can assist students to be expressive (Gilbert & Matthews, 2021; Matthews, 2021). Kimberley Foster's creation of dialogical objects (Fig. 5.2) introduced a tactile and material relation to arts interventions in the Earth Crisis (Sorhed, 2023). Additionally, Kevin Davidson and Tom Mansfield contributed innovative research in sound and movement (Fig. 5.3). We planned an assemblage of three arts methods for new ripples of creative mediation: (1) dialogical objects, (2) sound and movement and (3) writing and drawing. In this chapter I hope to show how arts practice research can be a space for happenings in empathic dialogue and making.

Situating the Eco-literature

Extractivist capitalism is currently drilling down into the Earth's precious resources, and pollution is shattering the atmosphere. The heat is literally on. Practice and literature that remind humanity of our responsibility to protect the Earth are supported by posthumanist and decolonial theory.

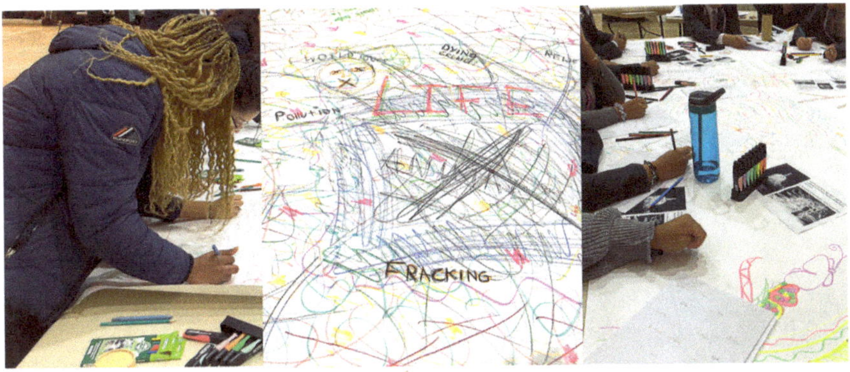

Fig. 5.1 Drawing and writing, with Miranda Matthews and Francis Gilbert

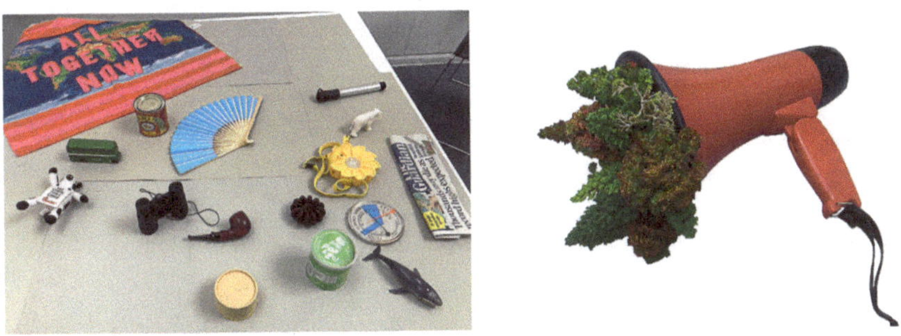

Fig. 5.2 Dialogical objects, curated by Bridget McKenzie and made by Kimberley Foster

The lives of today's children and generations to come are vital to us; however, human beings are not at the centre of the universe. Arts interventions that bring together local and international voices can connect diverse groups of people, with climate justice and reparation for all lifeforms.

For participatory arts methods with young people to take effect, there needs to be a gathering of ecological awareness, philosophical input, practice and pedagogical investigation. The causes and effects of the planetary crisis are all around us in urban locations. Our days are filled with the noise of polluting and extractivist machines. Our nights have no canopy of stars, only the orange glow of streetlamps. We live in areas of dense overcrowding, where huge properties are left empty by tycoons. Record

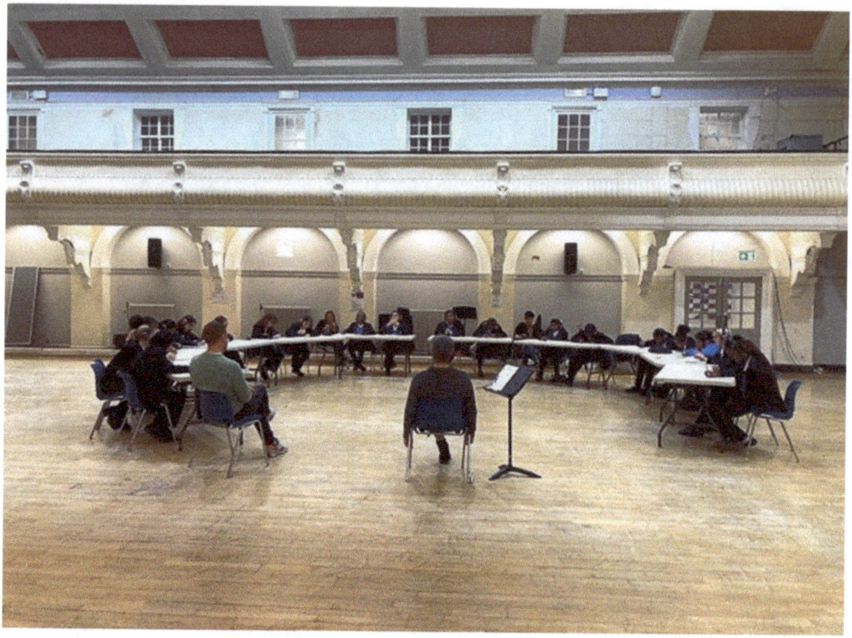

Fig. 5.3 Sound and movement, with Kevin Davidson and Tom Mansfield

temperatures in recent times have reduced grasslands to parched stubble, and wildlife populations are dwindling. Literature in the milieu of arts and educational studies in this context co-exists with cultural activism, in organisations such as Extinction Rebellion (Stammen & Meissner, 2022), Culture Declares Emergency (2023) and Climate Reframe (2023).

Posthumanist Momentum

Our approach is essentially posthumanist, in that we are exploring how to bring attention back to humanity's situation among other-than-human beings, and more-than-human environments. The Ecologies in Practice project gathered momentum with a relation to Deleuzian theory. Our collaborative activism began in the middle of things. The 'milieu' for action in education (Deleuze & Guattari, 2013) was intensified by happenings around COP26, the global convention on climate change in 2021, Extinction Rebellion's artivism (Stammen & Meissner, 2022) and school strikes started by Greta Thunberg who said, '[N]o one is too small to make a difference' (Thunberg, 2019). As an arts educator I wanted to find ways for youth expression of eco-anxiety around climate change to include schools.

Picture a group of practice researchers, young people and teachers experiencing a carousel of arts methods: with a Deleuzian perspective we are in the milieu of pedagogical action for transformation. With a view through Guattari (2008) we could be forming alternatives to the dependent consumerism of highly processed culture to support a ground swell of accurate information and active practice networking.

During the Ecologies in Practice research programme in 2021–2022, I reflected on how active peer groups could be seen as 'motivational assemblages', with different modes of engagement that support participatory action (Matthews, 2023). The energies brought together in CAL and CMUK research I hoped would help create motivational assemblages among young people, so that these active groups might continue to make empowered interventions.

Instead of generating competition and a hierarchy of achievements in creating practice, motivational assemblages work at the roots of possibility. Discursive arts practices can produce inclusive modes of human and environmental interaction that gather energy, in combined beneficial forces. In assemblage theory (Deleuze & Guattari, 2013), social and material elements gather to form dynamic compounds. The three arts methods and our team of practice researchers offered carefully complimentary, yet intently different socio-material dynamic compounds for young people to experience. Our intention was that the participating school students would be equipped with energising artivism when returning to school projects and arts practice in 'free time'.

To say that we worked pragmatically is not to say that we were defeatist: we celebrated the challenges that happened during the research. Our approach to Ecologies in Practice corresponds with Dennis Atkinson's relation to the 'pragmatics of the suddenly possible' (2018, p. 4), with Manning and Massumi's investigation of how practice 'generates the impossible' (Manning & Massumi, 2014, p. 127) and 'collective individuation in the milieu of expression' (ibid., p. 30). Parallels could be found in Isabelle Stengers' 'ecology of practice', defined as 'a tool for thinking through what is happening' (Stengers, 2005, p. 185). In planning, delivering and evaluating ecologies in practice workshops, the team tried to make a project map for navigating how to introduce local and global reparations for young people via the arts.

Reflecting on the need for the West/Global North to bring about reparation, CAL and CMUK researchers observe that climate injustice has had devastating effects on the Global Majority. It is said that indigenous people of colour are the stewards of the Earth, carefully nurturing the environments they live in without destroying biodiversity (Girard et al., 2022). Decolonial ecologies therefore need to be increasingly central to the ethical and theoretical mapping of practice research in the Earth Crisis.

Decolonial Ecosophy

To decentralise whiteness, Centre for Arts and Learning collaborates with artists of colour in ecological inclusive events. Artists such as Jennifer Steverson (2023) resituate historical knowledges in the hands of postcolonial peoples. The decolonial transgressive theories of authors such as bell hooks (2019) can also be read in relation to posthumanist theories of assemblage and affect to restore equality for marginalised groups (Matthews, 2023). Researchers need to gather understandings of how experiences of race, gender and class intersect in climate injustice. Patricia Hill Collins' (2019, p. 172) discussion of how arts practice contributes to intersectional 'creative social action' is relevant for young people protesting about climate change. Likewise, Stuart Hall's (2018) writings on the significance of arts practice in 'transition zones' have significance for young people of colour entering university spaces to make ecological artwork.

In addition the CAL and CMUK team have been delighted to make connections with decolonial activists such as Suzanne Dhaliwal and with innovative groups like Flow India (2023) and TheatrEX Bangladesh. Many young people from the Global Majority have taken part in CAL-CMUK workshops, often with forthright creative self-expression.

Indirect and Critical Pedagogies

To accompany posthumanist and decolonial approaches to practice research, there are also pedagogies to situate in empowering youth voice. My theorisation of teaching relates to critical pedagogy, as this focuses on empathy, raises consciousness and persists with working towards more lateral power relationships and sensory literacies in learning and teaching (Freire, 1996; hooks, 2019; Matthews, 2021). The interrelation of arts methods in the Ecologies in Practice workshops intended to create sensory literacies—among experiences of dialogic objects, sound and movement, drawing and writing. We hoped that embodied, verbal, haptic, aural, visual and olfactory experiences could bring together vibrant interpretations of Earth Crisis issues.

Considering the eco-anxiety that young people and many adults are experiencing, we began with indirect pedagogies that could potentially free affective forces bubbling under the surface. The research team wanted to encourage agency in creative practice, and aimed to do this by delivering arts methods without becoming didactic. Indirect pedagogies, as addressed by Herner Saeverot, steer away from instructional didacticism. They aim to challenge the presence of destructive influences in society, via strategies that do not impose a formulaic response. Saeverot writes that, 'Every human being should find out for themselves how they wish to exist and be in the world. Nevertheless, education may have a role in the students' existence and being in the world' (Saeverot, 2022, p. 2). Yet in these crisis conditions, education needs to include the distribution of information that will assist equitable humane and interspecies ways of being in the world. Ecologies in practice could be seen as ways of conveying and co-creating more equitable ontologies. They offer what Saeverot calls 'intermediaries' (p. 7) 'to reduce tensions' (p. 13) such as eco-anxieties, sometimes with humour and lightness.

In my view, indirect pedagogies create an extending leverage for topics of weight and intensity—such as the Earth Crisis, supporting creative decompression and de-toxification in learning experiences. Yet there are difficulties with being indirect, as I will discuss in this chapter, and implications for the tangible materiality of the urgent issues that are in focus. CAL and CMUK have formed a dialogical relation between indirect and critical pedagogies, aiming to arrive at collaborative arts practices for discovery and recovery.

Supporting the Methods

Participatory art methods need a combination of careful planning and flexible space for adaptation to what happens in practice. The CAL and CMUK Ecologies in Practice project relates to an embodied a/r/tography methodology (La Jevic & Springgay, 2008) that values equal elements of arts practice, research and teaching. We located and pursued questions emerging through practice research (Bulley & Şahin, 2021). A multimodal methodology (Kara, 2020) brought together the ecological and posthumanist interests of the research team, with synergies found among the range of arts processes we work with.

The first Ecologies in Practice workshop for school students activated the creative voices of young people who are often in the midst of conditioning expectations for behaviour at school. The second workshop was adapted for undergraduate students, who have additional spaces of adult decision-making in their learning environments. The third workshop returned to school students, aiming to meet our intentions to empower youth voice around Earth Crisis issues.

Research ethics were particularly at the fore: guardians gave their informed consent for young people aged 14–17; however, the students had also chosen to participate. Students aged 18 and above signed their own informed consent. Health and safety provisions were carefully planned to ensure participants had a secure environment for the workshops. Schools and students are not identified in this research to protect their anonymity. When gathering participants we contacted artist teachers who were already interested in practice research to increase reciprocity. The five participating schools were state-funded comprehensive schools and academies in London. Four of the five schools were mixed gender; one was a school for girls. In total, 87 students took part in three workshops.

The CAL and CMUK research team had to respond to the pragmatics of working with school students. Practice researchers wanted to expand into a whole day of art activities, but it was not possible to release school students for a full day as they were working towards exams. Teachers who took part in the workshops as school group leaders were given a guide that outlined each of the three arts methods and suggested follow up activities. This 'Group Leader's Guide' was later made available for more schools to download.

In addition to the experiences gathered during the workshops, in observations made by young people and practitioners, their creative texts, drawings and performances, there were focus groups immediately after the workshops. The focus groups enabled an additional reflective space, for empowering youth voice. Data from these reflective spaces is included in the next part of this chapter that will discuss the findings in a sequence of discovery.

Sharing Findings

The research team hope that some of our findings will be transferrable to learning practices that have not had the same affordances of collective practice research, within the current pressurised conditions of arts education in school environments and beyond. It is very rare for six arts educators to get to work together with young people in the same time and space. Since 2023 our arts methods and findings have been shared with different groups of educators, as for example at an international Ecologies in Practice Conference at Goldsmiths in 2023, and with the Artsmark community in the UK (2024). An evaluation of CAL-CMUK ecologies in practice research is presented here—locating where we felt it was possible to make interventions in the Earth Crisis.

Stage 1

The first stage of the Ecologies in Practice project was highly speculative: could we really gather students from three schools together, and enable their experiences of three arts methods in a university space? What would happen if our precise timings for a three-hour workshop were thrown awry? Would we all work together and support one another? Could all the young people be included equally? Might the information that we shared with young people, the resources and techniques be all too much for them? Would anything meaningful actually happen that could be taken away?

In addition to these concerns, the paradoxes of aiming for freedom of expression while communicating pedagogical practices and also conducting research could be discussed at greater length. We had anticipated the conditioning presences of teachers, but did not know how these presences would affect the students. On the one hand, they were more comfortable with teachers being there, yet some hesitated to speak—looking for a teacher's approval.

The young people communicated their excitement and enjoyment of the arts methods.

> All the lessons we have had, and everything about climate change, it's all been focused on numbers, and that doesn't resonate with me. So something more creative, it gets the cogs turning.

> There were no restrictions on any of today's activities, I really enjoyed them all. And I like the fact that all of the staff kept saying that we were free to do what we kind of felt was right. We could basically express our ideas however we wanted to.

They also shared their misgivings and expectations for change in our methods and the learning environments they encountered.

> Maybe encourage students to spend more time outdoors, in nature, or a project such as a community garden.

Here the parameters of being in an enclosed learning space and not in nature had become more noticeable to this young person, though we had intended to enable a sense of increased accessibility and belonging in a space of higher education.

Tensions around how the expressive artworks that were made in the workshops could be validated were also communicated. Art lessons in school often have a structured development of practical skills, with less space for embodiment and affect. Collective drawing was also challenging for some as they worried about the wastefulness of long sheets of paper—even though the paper was recycled.

> I didn't really like the chaos of it, and the large scale of it, just because it kind of felt like wasting paper. If you saw it out of context you wouldn't know what it is.

Expressions of feeling around the Earth Crisis were very moving, as they emerged in drawings, words and movement. Significantly, the location of affective and sometimes autoethnographic aspects when discussing causal factors of the crisis can enable a blending of known and unknown, and a mapping of this relatable information via artmaking.

As project leader, I found that the practitioners learned about each other and the relative effects of the interventions we were trying to create during the first workshop. We learned about our need to prioritise the voices of young people, more than the performative aspects of our voices in specific arts practices. We also learned that compromises had to be reached among the practitioners. For instance, I wanted to ask the young people to make free drawings about their responses to loss of biodiversity, showing images of work by a few different artists, and Francis Gilbert wanted to empower free writing in responses to a poem on the impacts of air pollution. It was too much for the young people to encounter before they moved to the next arts experience in the workshop, and the team's approach had to adapt.

In the focus groups the students were asked to think about what they would like to take back to develop at school. As their teacher was also in the group, their speech was highly conditioned, and they did not know what they would be allowed to do at school. In one group, the teacher stepped in to say how their project on climate change at school would relate to the Ecologies in Practice workshop. Therefore our questioning and problem-posing also had to change to increase expressive youth voice.

Stage 2

For young people to communicate well, arts educators need to communicate better with each other. To support all of the practice researchers to act in full awareness of each other's eco-pedagogies, I asked each pair to create a table of 'content, ethos, and references'—for increased understanding of our differences and interconnections. Kevin Davidson and Tom Mansfield began with an ethos of assisting agency, listening to meaning in the sounds of other-than-human animal voices. Bridget McKenzie and Kimberley Foster started with wearable, found and made objects, to encourage conversation and diverse interpretations of extractivism, dystopia and utopia with 'an eco-lens on things' (McKenzie, 2023). Francis Gilbert and I began with an ethos of encouraging participant voice, working with indirect pedagogies to enable nuances of affect and ambiguity in drawing and writing. I linked to the 'geomorphic' work of Caragh Thuring (Williamson, 2023, p. 33), an artist who expresses the forces of nature and of art (Atkinson, 2018). Francis connected to a poem about fracking called 'They Came for the Shale' by James Tierney (2017).

In the second workshop, we had a small group of seven undergraduates to work with. This group was tiny in comparison to the 53 school students in workshop 1. However, they were highly vocal and advanced in arts practice. With this small group, each pair of researchers was free at one point, to observe the others in their processes of delivery and communication. The researchers all had a space to take part in each other's activities. This was highly enjoyable, and we found that understanding of the participants' responses became much better informed.

Participants articulated both their challenges and experiences of expressive liberation, with some instances as follows.

> People have been able to be quite open and productive here, because we are not sure what you want.
>
> When we are all diagnosing these sounds [in Sound and Movement] it's kind of bizarre but everyone is smiling and laughing at each other, because we all have no idea of what's going on.
>
> The fact that there were three workshops was great because I think that people have been able to maybe use what they feel is their best way of communicating, in one or the other of the workshops.
>
> It's just really nice to do something about the climate that doesn't feature massive anxiety especially because I have signed up to climate talks or for talks about extinction. It's like, what's that called, 'Heading for Extinction' and it happens every Wednesday.

It became clear that the intangibility of indirect pedagogies enabled participants to be 'open and productive'; but the research needed a more strategic focus in the context of youth interventions in the Earth Crisis. Decompression of eco-anxieties in creative freedom of expression was not yet enabling the participants to feel as though they were able to meaningfully make an intervention—knowing 'what's going on'. The collective research approach had to adapt again to evoke increasingly meaningful interventions. Productive findings observed in our participants' responses include evidence of how arts practice aerates the complexities of the crisis, in breathing spaces of making and reflecting without 'massive anxiety', connecting with the world views of self and others.

Stage 3

A sustained responsibility to empower young voices can be a wonderful stimulus for transformational arts methods. In workshop 3, we encouraged the shift towards enabling protest and critical pedagogies, for increased empowerment that might carry beyond the immanence of the present day. By the time of our third workshop with school students, we had developed fully collaborative, manageable content. We had changed tack to focus on problem-posing protest and critical pedagogies. However, this significant shift still held traces of gentle, maieutic and tacit, indirect pedagogies.

Participants in workshop 3 were very positive about their interactions in all three arts methods, and more convinced that they could develop creative interventions of their own.

> Maybe it's helpful in trying to do something, because it's about things that you can contribute to.

> It makes it more fun and interactive instead of looking at something and just copying that down. We get to think about it, and talk about, and imagine all that stuff.

These comments indicated that we had created forms of interaction that were starting to empower young people to feel that they could make a creative intervention in the Earth Crisis. Young people were also increasingly confident in validating their own forms of self-expression: the focus on protest seemed to turn anxiety into collaborative creative action (Collins, 2019).

Conclusions

In this research we found that participatory arts methods for ecologies in practice are often working with the starting points for young people who are just beginning to find out about the effects of the Anthropocene, as they step into personal, local and global awareness. Small amounts of information, arts techniques and material foci are notably far more useful than an intense sudden influx of new 'things to know'. Next steps in the development of arts methods to increase vocal agency for young people could address questions that emerged in the middle—the 'milieu'—of our practice research. How could a greater elasticity of time for empowering learning interventions be created? Arts practitioners had to adapt to the pressured time spaces available to students aged 14–18, and the different pressures upon undergraduate students. The frustrations of little time for making art seemed to connect implicitly with the short amount of time we feel we have to make a difference.

The practice researchers also needed to find ways of sharing knowledge for synergies and consistency. To tackle eco-anxiety and enable empowerment we decided to put more attention on the causes of the Earth Crisis, such as capitalist extractivism and wasteful industrialisation, and less focus on impacts such as global warming. We found that this enabled more empowered 'artivism' (Stammen & Meissner, 2022). In the process of working together, the differences in content, ethos and reference points for three pairs of practice researchers became increasingly mutually supportive. Synergies gathered as we supported each other's eco-pedagogies and sustained reverberations of connectivity for young people.

For our multi-modal combination of arts methods to enable maximum opportunities for inclusive student voice, we had to factor in teacher voice as the participating teachers also wanted to talk and to share their opinions. With a critical pedagogy view, it was good practice for the teachers to share their experience, although at times teacher voice shaped the outlet of student voice.

Enjoyment of the moment, heightened imagination and the wondrous expressions of young people appeared to be really spontaneous—bringing new possibilities to the world. Yet there was meticulous attention to the sequencing of interactive processes, ecological materials and qualities of the environment, before these apparently spontaneous experiences of young participants. This motivational assemblage seemed to mirror the generative processes of the natural world behind the scenes of human activity.

It was clear that ecologies in practice brought in sensory expression that could make interventions in the Earth Crisis more fully dimensional and inclusive, beyond a reassertion of human power through scientific knowledges. This research intervention leaves us with the issue of how to increasingly validate the artworks and expressive voices of young people, as they seek to make reparative changes in their day-to-day experiences. One participating school continued to pursue a climate change related project, including a collaborative mural relating to bleached coral in polluted oceans. The research team hopes that the findings of this study will contribute to arts practice becoming integral to expressions of youth voice on climate awareness, as concerns for ecology become centralised in education. Ecologies in practice workshops have offered an extraordinary burst of activity; the young people's own time will tell if they are able to gather an evolution of attention to the planet, in the extended orbits of their arts practice.

Conflict of Interests The author declares there are no conflicts of interest. This study was funded by UKRI [grant number: AH/W010348/1].

References

Atkinson, D. (2018). *Art, disobedience and ethics: The adventure of pedagogy.* Palgrave Macmillan.

Boulianne, S., Mireille, L., & Ilkiw, D. (2020). School strike 4 climate: Social media and the international youth protest on climate change. *Media and Communication (Lisboa), 8*(2), 208–218.

Bulley, J., & Şahin, Ö. (2021). *Practice research - Report 1: What is practice research? and Report 2: How can practice research be shared?* Retrieved September 30, 2023, from https://bl.iro.bl.uk/concern/reports/b51c0f52-9801-49d9-9f00-cca89741091b

Climate Museum UK. (2021). *Listen and respond: Cultural learning in the Earth crisis.* Retrieved September 30, 2023, from https://climatemuseumuk.org/2021/05/25/listen-and-respond-cultural-learning-in-the-earth-crisis/

Climate Reframe. (2023). Retrieved September 30, 2023, from https://climatereframe.co.uk

Collins, P. H. (2019). *Intersectionality as critical social theory.* Duke University Press.

Culture Declares Emergency. (2023). Retrieved September 30, 2023, from https://www.culturedeclares.org

Deleuze, G., & Guattari, F. (2013). *A thousand plateaus.* Bloomsbury Academic.

Flow India. (2023). Retrieved September 30, 2023, from https://flowindia.com/

Freire, P. (1996). *Pedagogy of the oppressed.* Penguin.

Gilbert, F., & Matthews, M. (2021). Affective digital presence: How to free online writing and drawing? *Journal of Writing in Creative Practice, 14*(2), 209–237.

Girard, F., Hall, I., & Frison, C. (2022). *Biocultural rights, indigenous peoples and local communities: Protecting culture and environments.* Routledge.

Guattari, F. (2008). *The three ecologies.* Continuum.

Hall, S. (2018). *Familiar stranger: A life between two islands.* Penguin.

Hickman, C., Marks, E., Pinhkala, P., Clayton, S., Lewandowski, R. E., Mayall, E. E., Wray, B., Mellor, C., & van Susteran, L. (2021). Climate anxiety in children and young people and their beliefs about government responses to climate change: A global survey. *The Lancet Planetary Heealth.* Retrieved April 9, 2024, from https://www.thelancet.com/journals/lanplh/article/PIIS2542-5196(21)00278-3/fulltext

hooks, b. (2019). *Belonging: A culture of place.* Routledge.

jagodzinski, j. (Ed.). (2018). *Interrogating the anthropocene: Ecology, aesthetics, pedagogy and the future in question.* Palgrave Macmillan.

Kara, H. (2020). *Creative research methods.* Policy Press.

La Jevic, L., & Springgay, S. (2008). A/r/tography as an ethics of embodiment. *Qualitative Inquiry, 14*(1), 67–89.

Manning, E., & Massumi, B. (2014). *Thought in the act: Passages in the ecology of experience.* University of Minnesota Press.

Matthews, M. (2021). Sensory literacy and empowerment in art and design. *International Journal of Art and Design Education, 40*(3), 581–598.

Matthews, M. (2023). *Arts methods for the self-representation of undergraduate students: Sensory transitions into university cultures.* Routledge.

McKenzie, B. (2023). *Culture, adaptation, transition and resilience, climate museum UK*. Retrieved September 30, 2023, from https://climatemuseumuk.org/2023/07/12/culture-adaptation-transition-resilience/

Saeverot, H. (2022). *Indirect education: Exploring indirectness in teaching and research*. Routledge.

Sorhed. (2023) Retrieved September 30, 2023, from https://www.sorhed.com

Stammen, L., & Meissner, M. (2022). Social movements' transformative climate change communication: Extinction rebellion's artivism. *Social Movement Studies*, 1–20.

Stengers, I. (2005). Introductory notes on an ecology of practices. *Cultural Studies Review, 11*(1), 183–196.

Steverson, J. (2023). Retrieved September 30, 2023, from https://www.genevajean.com

Thunberg, G. (2019). *No one is too small to make a difference*. Penguin.

Tierney, J. (2017). *They came for the shale*. Retrieved September 30, 2023, from https://poetrysociety.org.uk/publications/i-am-the-universe-anthology/

Verlie, B. (2021). *Learning to live with climate change*. Routledge.

Williamson, B. (2023). Caragh thuring. In *Art monthly* (pp. 32–33). Britannia Art Publications Ltd.

CHAPTER 6

How Can We Help Young People Improve Their Local Environments? How Can They Become Agents of Change?

Francis Gilbert and Anna Stewart

Introduction

Young people derive great benefit from local green spaces, yet once they have grown out of the playgrounds often situated within them, many feel alienated from them (Van Aalst & Brands, 2021). This research project, created at Goldsmiths University of London, sought to re-connect young people with their nearby green spaces by encouraging them to use art, poetry, film and photography to engage with and research their local parks. A tutor at the university brought together eight Goldsmiths students from diverse degrees such as Psychology, Creative Writing and

F. Gilbert (✉)
Educational Studies, Goldsmiths, University of London, London, UK
e-mail: f.gilbert@gold.ac.uk

A. Stewart
i2Media Research Ltd., London, UK
e-mail: anna.stewart@i2mediaresearch.com

M. Matthews (ed.), *Ecologies in Practice and Learning: Arts Interventions in the Earth Crisis*, Palgrave Studies in Educational Futures, https://doi.org/10.1007/978-3-031-82338-1_6

Sociology and researchers to work with a local school and foster these ecological creative research methodologies. This led to the local school children meeting up with councillors, park management, park user groups, community health organisations and the police, thereby bringing multiple stakeholders together in order to improve the local park. As the project progressed, it ignited a commitment to both social and ecological justice in the young participants. This enthusiasm resonated with local councillors and park groups and the creative contributions of the young co-researchers (YCR) illuminated fresh perspectives and fostered lasting transformations within the park.

Research Questions and Rationale

This chapter seeks to answer these research questions:

> *How can academics, university students and teachers help young people improve their local environments? How can young people be assisted to become co-researchers into their local parks and become agents of change for them?*

To understand the purpose of the research, it is important to consider why research should be conducted into the connection between young people and parks. Put bluntly, our parks have a problem with young people. While parks in the United Kingdom—and many worldwide—cater for children aged 0–8 years with playgrounds, they frequently make older children feel unwelcome, particularly poorer young people who cannot bring play resources to the park. This is because young people struggle to find their own spaces and activities in parks, often feeling they are unfairly blamed for anti-social behaviour in such spaces (Van Aalst & Brands, 2021; Brown, 2013).

This chapter aims to explore ways in which creative research methodologies can be mobilised to help young people become researchers alongside academics, undergraduates and postgraduates. Its structure explains how these research methodologies were successful.

It offers a literature review of relevant research conducted into parks; the use of creative research methodologies applied to parks and young people; and discusses the methodologies and findings of the research project, arguing that these creative research methodologies have a significant impact on the experiences of young people and the ecologies they practice in parks.

Literature Review

The rationale of the research in discussion here is based on posthumanist, new materialist theories (Fox & Alldred, 2015) that are premised on the assumption that human and non-human entities can enter into interaction, generating affective forces (Gilbert & Matthews, 2021). These forces can stimulate participation, identification and new meanings, enabling new forms of connection through engaging with digital technologies (Toohey et al., 2015) and space/natural environments by utilising inquiries into places which involve critical holistic thinking about the environment (Honeyford & Watt, 2020). This study combines the two angles—digital technologies and natural environments—and investigates connections, bonding and the consequences for knowledge, self/environmental awareness-building in young co-researchers (YCR) whilst exploring urban green spaces and interacting with developing technology such as the 360 camera we used for this project. This camera provided unique perspectives such as aerial views and compressed geographies.

The posthumanist, new materialist methodology which we follow, and which will be outlined here perceives parks to be 'assemblages', the bringing together of the human and non-human: park users, trees, grass, paths, litter bins, etc. (Fox & Alldred, 2015). These park assemblages are often 'territorialised' (Deleuze et al., 2013) in popular discourses: that is, taken for granted, not interrogated, simply assumed to have an existence. The research aimed to 'deterritorialise' park assemblages, through the researchers' poems, photos, films and more traditional research tools such as surveys. The project sought to enable us to see the park through fresh eyes, to generate new 'affective flows', whereby the creative outputs would affect their creators and their audiences, generating new, productive 'lines of flight' (Fox & Alldred, 2015, p. 401, Gilbert & Matthews, 2021, p. 3) and new ways of thinking and caring for parks. The investigation aimed to be 'rhizomatic' seeking to position all researchers and subjects of the research on an equal footing, so facilitating new and ever-evolving connections (Fox & Alldred, 2015, p. 401; Gilbert & Matthews, 2021, p. 9).

Methods and Methodologies

The findings section discusses the difficulties and affordances of achieving new thinking about parks' ecologies. Recent decades have seen a growing movement towards arts-based research (Smith, 2009; Ingram, 2014) which utilises the creative arts to empower young people to be researchers, as opposed to research subjects. Using a methodology called photovoice (Ingram, 2014, p. 311) Ingram and her researchers helped teenage girls research their own attitudes towards citizenship, schooling and their bodies. They created montaged, digitally manipulated photographs of their feelings about the world and their relationship with it.

The Parklife project aimed to build upon such research by using similar creative approaches. Connecting with the British Academy and Students Organising for Sustainability UK (SOS-UK), Francis Gilbert brought together undergraduates and master's students from Goldsmiths with pupils from a nearby secondary school in an area of high social deprivation.

Some of the data for this chapter was generated by two one-hour semi-structured group interviews with the YCR in their school. They were joined by the teacher involved with the project, who also offered comments. All the students had parents' permission to be interviewed and recorded. Other data studied were the creative outputs produced by the YCR and their university mentors, that is, films, photos, pictures, poems, stories; field notes from the hour-long meetings between the YCR and their university mentors, teachers and other parks people such as: parks management teams, park user groups, community health organisers, local councillors, climate-emergency education facilitators; results from surveys and evaluations carried out during the project.

Findings

This section explores what our research uncovered, showing that new connections and technologies helped reconfigure the ecologies of the park.

Generating Affective Flows

The young people interviewed reported it was their creative work they found most enjoyable and that most motivated them. They also felt that this creative work led to some park improvements. One pupil reported:

> What I enjoyed the most was the creative aspect, and how we got to express things about how we want the park to change. And also the fact that we actually got to talk to people who could make a change. Like our voices were actually being heard. And that people might actually do something because we were speaking to the right people for the job.

It appears that it was the creative work which gave this pupil the most space to 'express things about how we want the park to change'. In other words, the fact that they were being asked to write poems, draw pictures, make films, etc. about the park 'affected' their relationship with the park: they began to see it in a different way. Their creative work provided them with some sense of agency; they began to internalise and reflect upon what the park meant to them (Gilbert & Matthews, 2021).

Poetry and Affective Flows

Let's look at this 'assemblage' poem a young co-researcher (YCR) and a student at Goldsmiths made together about the park (Fig. 6.1).

All of the YCR went on 'learning walks' around the park, and would then respond creatively. This poem was generated, in part, because a young man had tragically recently been murdered in the park. This tragedy provoked much discussion within the project (see the following section on safety issues). The YCR were also saddened and depressed by the litter and environmental degradation of the park when they observed it

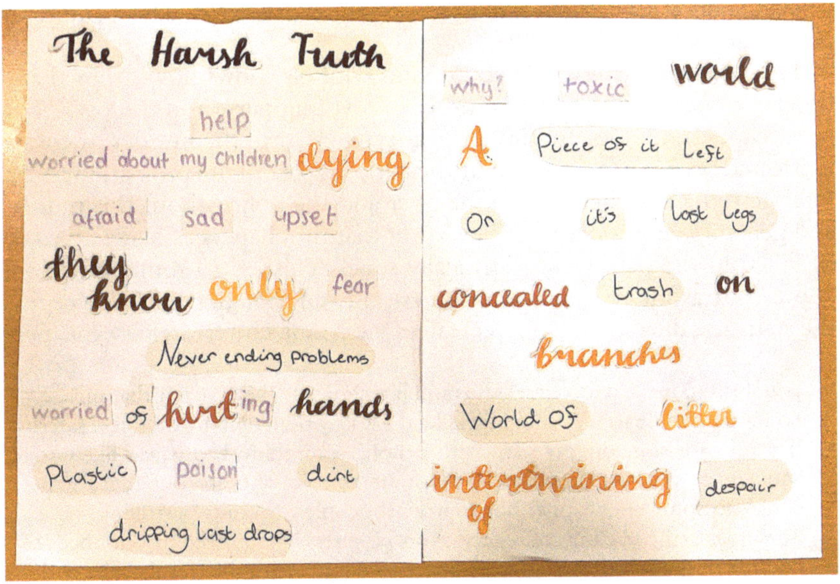

Fig. 6.1 A cut up poem about the park

closely and reflected upon it. Their learning walks generated an 'affective flow' (Fox & Alldred, 2015, p. 401, Gilbert & Matthews, 2021, p. 3) which led to the poem above. Drawing upon the words they handwrote, words cut out from a magazine, they wrote this emotional poem. The poem was read out by the young co-researcher to at the Advocacy Day where the young researchers shared their findings with a local councillor, a member of the park user groups, some members of the park management company, two representatives from a community health organisation and the council's schools climate emergency co-ordinator. The group were visibly moved by the poem, and the park's management representatives pledged to facilitate more litter picking in the park by involving the school. The councillor said he would seek to improve the lighting and safety of the park. In such a way we can see how a powerful 'affective flow' was generated by the creative research; the writing of the poem, in a way, led to action.

Safety Issues

Similarly, the YCR found it very helpful to draw and paint about the park. This generated some powerful emotional affective flows. This painting is particularly striking (Fig. 6.2).

Here the powerful use of colour and symbols shows how stunned and upset the pupils were by the murder of a young person in the park. The research methodology followed the affective flow of the murder: the ways it affected people's feelings about the park. As part of the project, the YCR interviewed a park user group member and the police about this murder and general safety concerns. They learnt that the chances of being attacked or killed in the park were statistically small. This led to some reflections about contrasting perceptions of the park. Even though the YCR accepted the statistics, they still did not feel safe. One young co-researcher explained:

> You want to make people feel safe even if technically they are not that unsafe. You want people to feel safe. I feel like if I was with people I would feel safer. If I am with someone, if I am with people, something bad is less likely to happen, especially being a girl. There are people who are out there who could do some crazy stuff. Being around people … adults patrolling. … it does make the feeling of safety more I guess announced pronounced? because of the fact that they are there … especially with the fact that the park is so big, and in certain places there is very little lighting, you can't really see

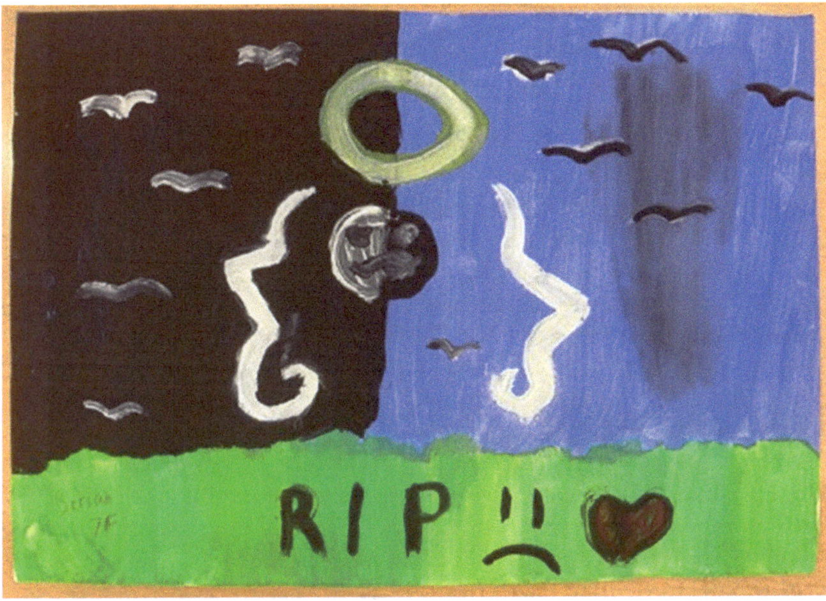

Fig. 6.2 A painting by a young co-researcher about a young man who was killed in the park

> what is in front of you. I have this thing where I am scared of what is in the dark, and if I can't see it really, really scares me, not being able to see my surroundings. Because the park is so open and is more like a field, open, not really much in between you.

As the young co-researcher notes, 'you want people to feel safe'. This became a vital element of the research, particularly after the YCR interviewed the police and pupils. In particular, the YCR conducted a survey about perceptions of safety in the park amongst pupils in January 2023.

The results from the survey found overall more students felt unsafe in the park than safe. 92% of Year 7s (total 95 pupils), 69% of Year 9s (total 64 pupils) and 65% of Year 11s (total 107 pupils) answered they felt unsafe in the park. This shows a clear trend that it is the younger students who feel most unsafe in the park. The underpass was reported where students felt most unsafe, followed by the grassy areas, paths, with the playground reported as the safest place.

This research led to the young co-researchers devising an interactive assembly for the school on perceptions of safety in the park, where they used the quiz format of UK television programme 'Who Wants to be A Millionaire' to question students about whether they felt safe in the park and where they felt least safe. Thus, they presented the results of their survey in a fun way by getting their audience to predict what the results might be, and then revealing them. They then followed up the revelation of the results with some statistics about crime connected with the park which indicate that crime is relatively low in the area (Metropolitan Police, 2022). The research was presented ethically in that it was not mobilised to shock or scare the school students, but rather to inform them, and address issues connected with perceptions and reality of the park. This research aimed to create a form of community ecology which reconnected young people with the park.

The young co-researchers also set about revivifying an old initiative whereby local shopkeepers, businesses and organisations advertised the fact that they were 'safe' spaces where people could go if they were feeling unsafe (Axis Foundation, 2018). The people running these safe spaces were re-interviewed and got 'back on board', and then posters were produced which advertised these safe spaces. Further research indicated that children in the school felt safer because of this poster campaign and the revitalising of the publicity about safe spaces.

New Technologies Providing Lines of Flight

There is, therefore, research evidence (Axis Foundation, 2018) to show that creative work can make young people feel safer in certain spaces. Certainly, the creative work that the young people carried out with digital technologies helped them feel more at home in the park. It also helped them to keep safe in the park by sharing accurate information and awareness. The university students on the project helped the YCR film and photograph the park using a '360' camera. This piece of technology can provide new perspectives by capturing a circular picture of its environment—front, behind and from the sides—a '360' degrees shot of an area. This creative work provided insights into the ecology of the park. Here is one still from the film which the students made of a small animal's view of the park (Fig. 6.3):

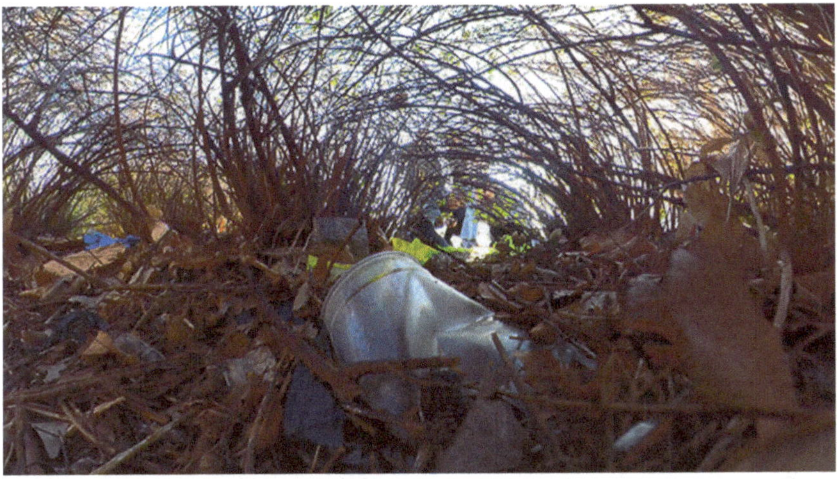

Fig. 6.3 Picturing an animal's perspective of the litter in the park

The YCR and the university students who made this film believed it served a vital purpose: it provided a vital insight into how animals might perceive the park. One young co-researcher said:

> You can say all you like about the environment and how litter is bad for it, and you can show lots of statistics and stuff about how bad life is for animals, but when you watch this film, when you look at this photograph, you can feel the pain of the animal, you know what I mean? For an animal it's a world of danger created by people's carelessness.

The new technology (at the time of writing) of the 360 camera afforded a 'line of flight' (Fox & Alldred, 2015, p. 401, Gilbert & Matthews, 2021, p. 3), that provided a new insight into the ecology of the park because it viscerally shows an animal's perspective. The research process which was facilitated substantially by the 360 camera is 'posthuman' to the extent that it affords perceptions of the non-human, taking into what Deleuze and Guattari call the 'root' and 'burrow' of knowledge (Deleuze et al., 2013) Similarly, another photograph taken by the 360 camera shows a bird's eye perspective of the park and school (Gilbert, 2022) which one student commented upon:

> What I like about the photograph is that you get the whole of the park in it. It's a rectangular park but the camera squashes it all into a circle, and you get sense of how it's quite small really. The sky is much bigger, and it's surrounded by buildings, but it's beautiful somehow. You feel like a bird looking at it, floating high above it. It makes you think how precious it is, how delicate somehow.

While the technology of the camera can powerfully convey the horror of an animal's perspective of litter on the ground, it can also provide us with a deep sense of beauty. The young co-researcher quoted above felt that it showed how 'precious' and 'delicate' the park is: this insight into its ecology was created by the camera's ability to capture a bird's eye view.

The process of using the 360 camera to record video and photographs is explorative by nature and using it can create new lines of flight. The viewer can see in real time a depiction of what the camera is shooting but there are numerous possibilities for shaping the final view of the photo/video into something completely different from the original shot with the use of editing tools. One young co-researcher saw a bush with litter and thought to put the camera inside, not knowing how it may turn out. It was not until the editing stage that it became apparent this could create a non-human perspective, thus becoming 'the picture of an animal's perspective of the litter in the park'. This generated new affective flows such as thoughts and further research ideas to be integrated into the project.

The creative research process here did not necessarily lead to 'hard' actions in any tangible sense (i.e. pledges by the adults at the Advocacy Day) but it did 'affect' all the research participants, many of whom reported being deeply affected by the photos and film of the 360 camera. One of the attendees at the advocacy day, a member of a community group, said that she felt very 'moved' by the film, and its offering of different perspectives of the park. 'It made me see it in a light I had never seen before', she said. 'It made me want to treasure it more.'

Rhizomatic Connections: Connecting the Human and Non-human

The organisation of the project was innovative in that it brought together many different stakeholders connected with the park, but not in a hierarchical fashion, rather in a rhizomatic way. There were no clear 'chains of command' like you might find in a hierarchically structured research project, with 'supposed' experts being in charge of less 'expert' researchers. The research project was not 'arboreal' in this sense: it did not have a tree-like structure with so-called important people being the trunk of the tree and the 'lesser' people being its branches and leaves. Rather it was 'rhizomatic' in that processes of connections were established and networks created which were non-hierarchical.

One young co-researcher said:

> I really liked working with the university students because it felt like they were adults, but they were also our equals. They really listened to us.

This notion of equality is central to the idea of rhizomatic connections: the YCR reported that they felt their knowledge of the park was as valued as the university students' more academic knowledge. This created a reciprocal exchange of knowledge and ideas.

This said, the YCR valued the university's academic knowledge. Another young co-researcher said:

> I thought the research process was really enlightening. I feel like the university students really helped, especially since they were university students, they knew a lot more than us, and that they helped us learn, and they helped with the research, because as university students they've probably researched a lot more things. They've probably made a lot of essays, talked to a lot of people about certain issues. They had a lot of experience, I feel like they were a bit like mentors in order to help through the process.

The comment here about the university students knowing 'a lot' more than the YCR needs to be taken in context: what was meant here was academic knowledge. This is an important point because the YCR really valued how they learnt about research processes from the university students and the other adult helpers on the project. The YCR felt safely 'held' by the university students: they appreciated the university students' 'experience', their ability to write 'essays', and their discussions with other people.

Switching to the psychology postgraduate university student's perspective, the research provided lines of flight in the form of a new research assemblage. Contrary to the standard psychological research methods of forming and testing hypotheses, there was no formal hypothesis for this research. Such research led the student to come out of their 'comfort zone' and approach the research process in a different light from taught studies. This research process meant forming research questions along the way, updating and creating rhizomatic connections not usually encountered within experimental psychology settings where variables need to be controlled and procedures planned out. This also led the university student to be open to developments, to form connections between the human and non-human and to think of new ways of integrating interdisciplinary practices into future research projects.

Additionally, the university student noted the adaptability displayed by the YCR throughout each stage of the project, and how they approached creative and traditional methods of research alongside the more serious topics such as safety and crime within the park. As time passed, a noticeable transformation took place, with the YCR growing increasingly self-assured in articulating their ideas and with help from the university student their academic knowledge base grew. This in turn helped the university student's sense of confidence and direction for the project.

In respect to the young co-researchers' embodied academic knowledge, this led them to being able to co-construct some important short and long-term goals for the project that were split into three themes: safety, litter and engaging young people. The young people presented these goals in a chart. Students aimed to achieve things like the setting up litter picking and gardening groups, better lighting, the establishment of sports' equipment, community gardens, water fountains and more sports facilities.

This goal setting led to more rhizomatic connections (Deleuze et al., 2013; Gilbert & Matthews, 2021). Their goals and the chart were presented to the local councillor, community group representatives, the parks' management team and the environmental educators on the Advocacy Day, and acted as powerful incentive for these people to act upon what they learnt from the research presented. During the Advocacy Meeting, these various stakeholders 'adopted' different aspects of these goals. The parks' management team took on the job of assisting with improving the littering of the park and its lighting, the local councillor promised to address wider safety concerns and the community group said it would work with the parks' management to set up a community garden. During the next year, all the pledges in the chart were being acted upon and achieved.

There were not only human rhizomatic connections but also artistic, creative and literary connections made. There were some interesting rhizomatic connections between the creative and more formal elements of the research: the pictures, collages, films, poems (creative elements) 'kick-started' and then intersected with more 'traditional' research outputs such as surveys, interviews and the establishment of short-term and long-term research goals. This intersection between the creative and the more established research strategies is nicely captured in one YCR's observations:

> I enjoyed the same things as everyone else, and also I enjoyed working with the 360 camera, something I've never done before. It was all about learning new things. I liked how when we were doing the creative research, we were focused upon what we wanted to change. And then we were doing the research like surveys, and we were looking at people's views on what they wanted to change, and then we were doing research to make that change happen. We gathered the information about what the situation is now, and how we can change it to make it what we want it to be. We did a storyboard of the park, and what the problems were, like the bins all overflowed, and then we went out, and we actually saw what the problems were. And then we devised solutions, to do things like put in more bins.

This young co-researcher saw that the creative research was a good way of understanding what their own attitudes and desires for the park were, while the more formal research strategies such as devising surveys helped them understand what other people's attitudes were. They also understood it was important to research how they might make change happen.

Conclusions

The research process followed and tracked a complex affective flow, which included the elements of people feeling unsafe when alone or in the dark in the park; the murder of a young person; lack of lighting or visible responsible adults; surveying pupils about perceptions of safety; educating pupils about perceptions of safety and the reality of crime in the park; the revivifying of readvertising/promoting safe spaces and the desire for more facilities in the park: sports equipment, bike lanes, community gardens.

All of these elements were expressed in the young people's creative work, which in itself instituted more affective flows; the creative outputs affected adults in power on the Advocacy Day, and this led to change for the better in the park. The Parklife project's innovative positioning of creative responses to the park as research created new affective flows which led to park improvements in terms of its upkeep, ecology and publicity.

The poems, paintings, collages and films that the young people made enabled them to better understand the park: their feelings towards it, its ecology, layout and geography, and its place within the wider culture of the borough. These creative outputs generated 'lines of flight' (Gilbert & Matthews, 2021, p. 5); they fostered new thinking about the park, seeing it as an ecological being in itself, vulnerable and in need of care. These lines of flight included the researchers considering how it might be improved, leading to the YCR speaking to significant adults connected to the park. As a result, a new water fountain, better lighting, better litter picking schedules and a community garden were established in the park. In other words, the affective flows had tangible affects.

This creative research increased the connections between diverse communities: researchers, university students, teachers, pupils, parks' management, councillors and other community members. Students at the university learnt to communicate and mentor the young co-researchers (YCR) in a motivating fashion, offering a powerful psychological space for discussion about the park. It proved a great starting point for more traditional types of research too, with surveys, questionnaires, interviews and literature reviews 'sprouting' out of the creative research in a rhizomatic, organic fashion. The YCR and university students were able to learn from each other. The YCR learnt how to use new innovative creative research methods as a means of voicing their opinions about their local park and speaking about issues that they may not have given much thought to before. As for the university students, they were able to disseminate their practices to the YCR, developing ideas along the way and learning what both motivated and challenged them, and the YCR, as the project progressed. Community and ecological connections, in all their rhizomatic complexity, were created and strengthened.

Conflict of Interests (2) Francis Gilbert has a received research grants from Goldsmiths and the British Academy for this project.

(4) The authors have no conflicts of interest to declare that are relevant to the content of this chapter.

References

Axis Foundation. (2018, June 22). *For Jimmy, creating safer spaces and communities*. Retrieved January 23, 2023, from https://axisfoundation.org/for-jimmy/

Brown, D. M. (2013). Young people, anti-social behaviour and public space: The role of community wardens in policing the 'ASBO generation'. *Urban Studies, 50*(3), 538–555.

Deleuze, G., Guattari, F., & Massumi, B. (2013). *A thousand plateaus: Capitalism and schizophrenia.* Bloomsbury Academic.

Fox, N. J., & Alldred, P. (2015). New materialist social inquiry: Designs, methods and the research-assemblage. *International Journal of Social Research Methodology, 18*(4), 399–414.

Gilbert, F. (2022). *How can we help young people improve their local environments? How can they become agents of change?* Retrieved April 15, 2024, from https://sites.gold.ac.uk/educational-studies-blog/how-can-we-help-young-people-improve-their-local-environments-how-can-they-become-agents-of-change/

Gilbert, F., & Matthews, M. (2021). Affective digital presence: How to free online writing and drawing? *Journal of Writing in Creative Practice, 14*(2), 209–237.

Honeyford, M. A., & Watt, J. (2020). Practising place-thought and engaging in critical place inquiry in a public park: Listening, letting go, and unsettling our writing pedagogies. In K. Toohey, S. Smythe, D. Dagenais, & M. Forte (Eds.), *Transforming language and literacy education* (pp. 32–53). Routledge.

Ingram, L. A. (2014). Re-imagining roles: Using collaborative and creative research methodologies to explore girls' perspectives on gender, citizenship and schooling. *Educational Action Research, 22*(3), 306–324.

Metropolitan Police. (2022). *New Cross crime statistics.* Retrieved January 24, 2023, from https://www.met.police.uk/a/your-area/met/lewisham/new-cross/

Smith, H. (Ed.). (2009). *Practice-led research, research-led practice in the creative arts.* Edinburgh University Press.

Toohey, K., Dagenais, D., Fodor, A., Hof, L., Nuñez, O., Singh, A., & Schulze, L. (2015). "That sounds so cooool": Entanglements of children, digital tools, and literacy practices. *TESOL Quarterly, 49*(3), 461–485.

Van Aalst, I., & Brands, J. (2021). Young people: Being apart, together in an urban park. *Journal of Urbanism: International Research on Placemaking and Urban Sustainability, 14*(1), 1–17.

CHAPTER 7

Ecologies of CARE: Reflecting on an Erasmus+ Research Project

Emese Hall

Introduction

Visual art education in new times: Connecting Art with REal life issues (CARE) was an Erasmus+ research project that ran from 2019 to 2022. The research partners were based in universities in Cyprus, Malta, Greece, and England, along with critical friends in Germany, representing the International Society for Education through Art (InSEA). Project CARE's principal aim was to enhance visual art education (hereafter 'art education') with education for sustainable development (hereafter 'sustainability education') within primary education (ages 5–11). Art education has a central role in assisting young people to make sense of the Earth Crisis, not least because many contemporary artists address the Earth Crisis in their work (Hall, 2023). One aspect of the project's originality was that, before the start of the study, most research into the potential synergy between visual art education and education for sustainable development was conducted

E. Hall (✉)
School of Education, University of Exeter, Exeter, UK
e-mail: emese.hall@exeter.ac.uk

M. Matthews (ed.), *Ecologies in Practice and Learning: Arts Interventions in the Earth Crisis*, Palgrave Studies in Educational Futures, https://doi.org/10.1007/978-3-031-82338-1_7

with older students (e.g., Hasio & Crane, 2014) and rarely involved trainee teachers (Inwood & Kennedy, 2020). The study's participants were a mixture of pre-service and more experienced teachers in the four participating countries. To indicate scale and impact, CARE involved nearly 120 teachers in training events and nearly 1300 children in learning activities. Further, approximately 2700 people engaged with various research dissemination events in-person and online over the course of the project.

The following definition of sustainability education created by the Maltese project partners threaded through all aspects of the study, as both children and adults were viewed as learners and active participants:

> [It] enables learners to develop the knowledge, skills, attitudes and values required to become active participants, individually and collectively, in decision-making processes, both at local and global levels that will improve the quality of life of present and future generations. (Vella et al., 2022, p. 10)

Relevant knowledge, skills, attitudes, and values were informed by UNESCO's (2017) Global Goals, also known as the Sustainable Development Goals (Fig. 7.1), which served as a reference framework for CARE's interventional content. As detailed in Fig. 7.1, the 17 Global Goals encompass multiple economic, social, environmental, and cultural issues. According to UNESCO (2023b), 'Culture and education are the backbone of sustainable development. Together, they shape who we are

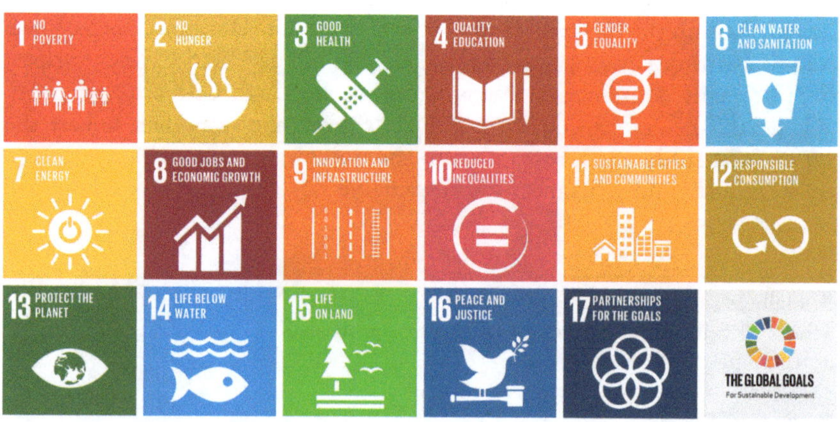

Fig. 7.1 The 17 UNESCO global goals

and who we will become'. However, the cultural dimension of sustainability is not always acknowledged. For example, the Department for Education in England, in common with many other organisations, only explicitly lists these issues as environmental, social, or economic (e.g., DfE, 2022). In project CARE, with art education at its core, the cultural dimension was important to foreground. Notably, the Earth Crisis is a catalyst for contemporary artists presenting potential solutions to ecological issues rather than merely creating representations of the issues themselves (Fowkes & Fowkes, 2022). The project used a collection of more than 60 contemporary artworks, drawn from each partner country, linked to the Global Goals as springboards for learning in and through art—a clear rationale for this is shared here by the CARE Principal Investigator, Victoria Pavlou (2022, p. 7):

> Contemporary artists are inspired by reality and use art as a means to advocate for a reconstruction of society, to discuss issues that will lead to sustainable societies and sustainable living. In the educational context, this proximity to the real experience has a significant transformative impact on the learner.

The researchers approached a diverse range of artists to gain permission to use their artworks in digital form. Further, some of the artists' work was viewed in real life by the children (e.g., amongst the participants in Greece via workshops), increasing its relatability and transformational impact on learning.

Key objectives determined the structure of the study, the methodology of which was based on action-research approaches (e.g., Baumfield et al., 2012): researching teachers' competences and needs in art education and sustainability education; designing and delivering training programmes about art education and the Earth Crisis within each partner country; supporting participants to teach and evaluate schemes of work in schools; creating professional learning communities (PLCs); researching the training programmes, the impact of the schemes of work, and the benefits and challenges of the PLCs. My role was Principal Investigator for the English team, leading on the PLCs. Professional learning is a particular area of my research interest and expertise, drawing on nearly 20 years of experience within teacher education, and I fuse this with my interest and expertise in art education, especially art education for the Earth Crisis (e.g., Hall, 2023).

Having offered a short overview of Project CARE, the chapter next discusses applicable interpretations of ecology and the how these manifested in two separate aspects of the study. I should also clarify that the reflections presented here are mostly new, prompted by revisiting the research with an ecological lens; this has been an insightful exercise.

Defining Ecology

Despite being a frequently used term, ecology is not always given much attention in documents about sustainability, and sustainability education. For example, in The Stockholm Environmental Institute's report, 'Charting a youth vision for a just and sustainable future' (Aggarwal et al., 2022), consisting of 70 pages, there is only one mention of ecology, and this is within the term 'agroecology'. To give another instance, in the Department for Education's (DfE) (2022) policy, 'Sustainability and climate change: A strategy for the education and children's services systems', an online publication of nearly 11,000 words, there is zero mention of 'ecology'. One wonders why this might be. A possible explanation is that the term ecology is often used as a synonym for environment. For example, we may talk broadly about 'the ecological crisis' and 'the environmental crisis' and mean exactly the same thing. If we search again within The Stockholm Environmental Institute report (Aggarwal et al., 2022), there are 86 mentions of environment (excluding within the reference list). Similarly, to repeat the same exercise with the aforementioned DfE (2022) strategy, we find 53 mentions. Although it is possible to make a distinction between ecology and environment, this distinction is not always clear cut and there are overlaps in my own use of these terms as well as in the writing of other CARE researchers (e.g., Vella et al., 2022). Importantly, I will now elaborate on how ecology/ecologies are being specifically applied within this chapter.

Within the Oxford English Dictionary, ecology is defined as: 'The branch of biology that deals with the relationships between living organisms and their environment. Also: the relationships themselves, esp. those of a specified organism' (OED, n.d.-a, Definition 1a). With reference to 'living organisms', I argue there is a tendency to think of ourselves (humans) as separate from other living organisms but, certainly, from a biological perspective this is erroneous (Lent, 2021). Expanding on this, Stibbe (2021, p. 4) argues that,

The danger of focusing on difference is that it can obscure some of the important things that humans and other animals have in common: having emotions, being embodied, bonding socially with others, and, most importantly, being dependent on other species and the environment for our continued survival.

Relationship with environment is of particular relevance to sustainability education, as reflected in one Oxford English Dictionary definition of environment: 'The natural world or physical surroundings in general, either as a whole or within a particular geographical area, esp. as affected by human activity' (OED, n.d.-b, Definition 2d). The reference to human activity here connects with the interrelated issues of human-made climate change and unsustainable use of natural resources that UNESCO's (2017) Global Goals seek to address. A clear distinction also exists in the UNESCO framing of environment between life below water and life on land.

A further Oxford English Dictionary definition of ecology offers a sociological perspective: 'The study of the relationships between people, social groups, and their environment' (OED, n.d.-b, Definition 1b). This brings to mind Bronfenbrenner's (1979) Bioecological Model of Human Development, illustrating the possible connections between different systems within which any individual interacts. There are many versions of this model, but all have a nested format indicating the systems hierarchy, with the chronosystem (time) being the largest and most universal. Interestingly, Stanger (2011, p. 169) critiques the 'eco' aspect of Bronfenbrenner's model because the environments to which Bronfenbrenner refers are mainly 'human, cultural, and social environments' and less so natural environments. However, for the purpose of my chapter this interpretation of an ecosystem is fitting because within Project CARE, as a social science research study, the participants' interpersonal relationships were of particular interest in terms of fostering desirable knowledge, skills, attitudes, and values of relevance to both sustainability education and art education. The project sought to promote these relationships via the facilitation of PLCs—bringing together teachers with shared interests to create a community with a common sense of purpose.

To briefly summarise: the two definitions of ecology shared above, the biological and the sociological, correspond to the proceeding discussion of Project CARE within this chapter: firstly, from the ecological consideration of respecting and protecting the natural environment—of which we are a part—and, secondly, concerning the 'ecologies of practice' (e.g., Stengers, 2005) evident in PLCs.

A third understanding of ecology also applies, related to the sociological understanding of ecology and Bronfenbrenner's (1979) modelling. As a study with an action research-informed methodology, some of the CARE ecologies of practice were created by the research itself. For example, the PLCs arose because the research design required their establishment. Furthermore, with its interdisciplinary focus combining art education and sustainability education, a new 'practice architecture' (Kemmis, 2022) was developed as the study progressed. As Kemmis states, 'When multiple practices are connecting with one another in a site, and if enough of these relationships of mutual interdependence and entanglement are formed, then an ecology of practices may form and begin to develop and evolve' (2022, p. 119). To exemplify this relationship, within Fig. 7.2, 'Practice AE' signifies art education, and 'Practice SE' signifies sustainability education. From an interpretive ontological perspective, the particular fusion of art education and sustainability education (AE*SE in the diagram) was unique to the project. Conceptually, these distinct practices of art education and sustainability education were similar across the four research partners due to shared understandings underpinned by the study's broad framing (the exosystem). However, there were differences in these practices too, as highlighted in the first stage of the project when, for example, country-specific educational policies concerning art education and

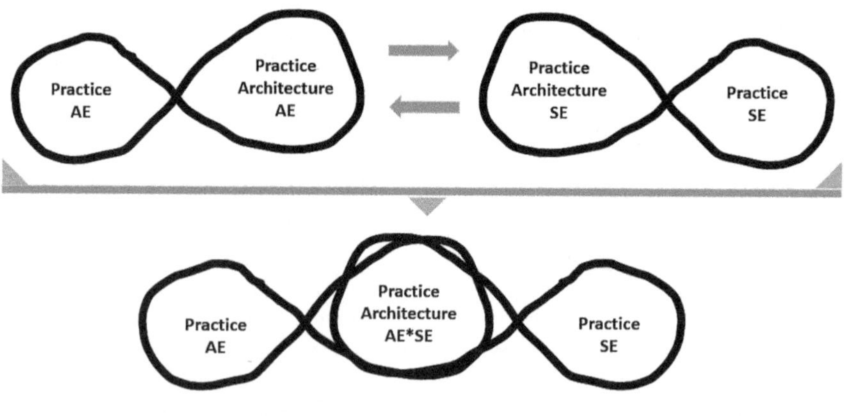

Fig. 7.2 Two practices becoming enmeshed through shared practice architectures (adapted from Kemmis, 2022)

sustainability education were analysed (Pavlou, 2022). Further, more nuanced differences emerged as the study progressed and data were gathered from the research participants based on their lived experiences.

There is another relevant observation regarding this visualisation that I would like to share. In Kemmis' (2022) accompanying explanation of the 'entangled' relationship between ecologies of practice shown in Fig. 7.2, rather than the disciplines/fields of knowledge that I have assigned, the practice belongs to a person, with the teacher positioned on the left of the model and the student on the right, and the new practice architecture arises via teacher-student interactions. The interesting slant on this apparently clear-cut dichotomy is that in the context of Project CARE the participants, as professional learners, had both teacher and student roles within the study. A key distinction is that, unlike when the practices of two different people combine, the teacher as student (in a dual role) will be familiar with their own ways of thinking and doing, even though new insights may be gained over time. In particular, the final phase of the study highlighted these new insights via the teachers' reflections on the case studies of the art units in schools.

The next part of the discussion will consider the first interpretation of ecology with Project CARE, as highlighted above. Although listed as 'first', no hierarchy is intended.

Ecologies Within the Project's Knowledge Base

The first angle on ecology in Project CARE concerns the ecological consideration of respecting and protecting the natural environment—an important component of sustainability education. I refer to this as the ecologies within the project's knowledge base, recognising that this knowledge was by no means always pre-determined and static but, concurring with socio-cultural learning theories (e.g., Kozulin, 2023), also actively created by the participant teachers through their interaction with programme content and other people—including children. Neatly, this perspective aligns with Kemmis' (2022) original explanation of the creation of new practice architectures and my reinterpretation shown in Fig. 7.2.

Ecologies within the CARE's knowledge base were evident in various parts of the research and most notably the training programme content, which was mainly hosted online due to COVID-19 restrictions affecting social mixing. Learning intentions for the partners' local training programmes were informed by the expert guidance provided by the Maltese research team, who led on this part of the research. For example, a key organisational framework was offered in the use of six 'Big Ideas', one of which was ecological literacy (Vella et al., 2022). Where ecological literacy was discussed in the CARE training guide, it was noted that:

> An ecologically literate society would be a sustainable society that avoids destroying the natural environment on which they depend. Since the survival of humanity depends on nature, ecological literacy must become a critical skill for all and included as part of education at all levels from primary to professional development (Capra 2008). (Vella et al., 2022, p. 24)

As previously stated, UNESCO's (2017) Global Goals can be categorised into economic, social, environmental, and cultural issues—bearing in mind the interconnectivity between these systems (Bronfenbrenner, 1979). For the present definition of ecology regarding respecting and protecting the natural environment, these are the six environmental Global Goals (GGs):

- GG6: Clean Water and Sanitation
- GG7: Affordable and Clean Energy
- GG12: Responsible Consumption and Production
- GG13: Climate Action
- GG14: Life Below Water
- GG15: Life on Land

Given the scope of this list, I will emphasise, as I have done elsewhere (Hall, 2023), that in talking of the climate crisis or climate education we are unhelpfully neglecting the other environmental issues that are connected to climate concerns. Yes, global warming is a huge concern but so too, and perhaps more so, is environmental destruction—and this all leads to global warming (Hickel, 2022). Ecological literacy, especially when developed through learning in the arts, enables the interconnectedness between activities and events, both human-made and natural, to be recognised and understood.

The training programme in England specified three learning intentions for the designated topic on ecological literacy:

- To understand the meaning and relevance of 'ecological literacy'
- To critically discuss different examples of ecologically themed art, using pre-determined frameworks of analysis
- To consider the possibilities of enhancing ecological literacy through art education

When the participants designed their units of work to teach in schools, in terms of Big Ideas, ecological literacy was the most popular theme amongst the trainee teachers in England. For example, one trainee teacher, whose school was in a coastal location, chose to focus on ocean-centred issues which strongly connected to GG14: Life Below Water. Here the neighbourhood of the mesosystem was clearly influential (Bronfenbrenner, 1979). An emphasis on ecological concerns was also evident across the many units of work designed and delivered by CARE participants in the other three partner countries, despite social concerns also being addressed (Ioannidou & Mitakidou, 2022).

The case studies of the art units in schools included some examples of the children's artwork. In Fig. 7.3 you can see the outcome of a school-based competition to create an eco-themed mural in Cyprus. The winning drawing was painted by a group of children, working collaboratively. The specific explanation of the design is not offered in the CARE report (Ioannidou & Mitakidou, 2022) so it is open to interpretation. However, one can view the cupped hands as universally recognisable caring gesture towards the Earth, and the butterfly might be a viewed as a concerned onlooker, thinking or saying: 'I might be fragile, but so too is our planet'. A slightly different approach to the painted mural about environmental issues is the use of natural materials to create artworks and many of CARE's school-based activities made use of these. For example, simple small stick sculptures were made by five and six-year olds in one of the English schools. Their (trainee) teacher planned this as part of the first of a series of lessons where they could begin to appreciate the affordance of natural materials in order to make informed choices when working with them throughout the project, which was titled 'Nature's Great Makers' (linked to GG15: Life on Land). This focus very much promoted children's knowledge of the web of life (Lent, 2021).

For Peacock (2004, p. 3), teaching ecological literacy can be highly motivating but one must also bear in mind that 'We have to help children to *want* to learn more about their world, and to use their learning to bring

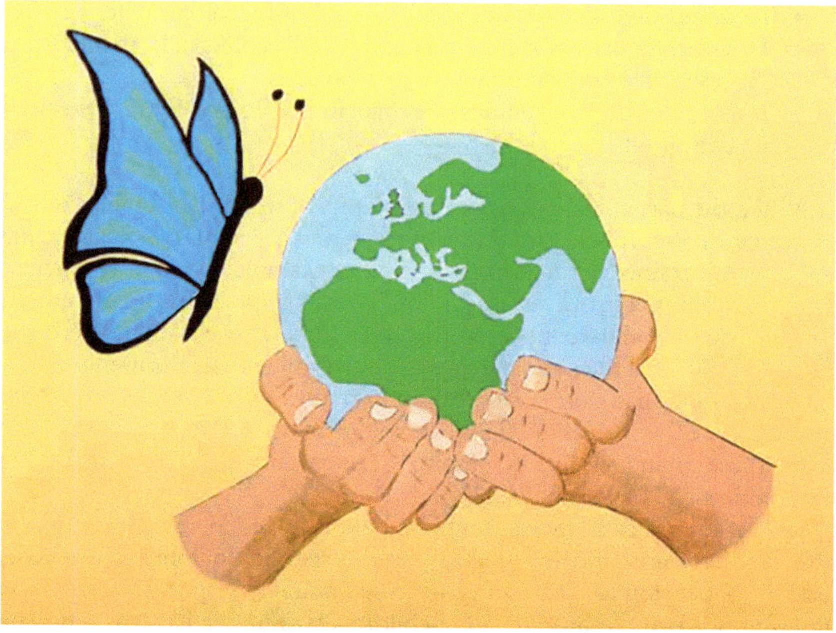

Fig. 7.3 Cypriot children's collaborative mural design (Ioannidou & Mitakidou, 2022)

about change for the better'. The evidence from the CARE case studies of the art units in schools showed clear evidence that learners wanted to learn more.

Having offered some discussion about a biological understanding of ecology within Project CARE, the next section addresses a sociological understanding.

Ecologies Within the Professional Learning Communities

The second angle on ecology in Project CARE concerns the ecologies of practice evident in the PLCs that were created formally and informally within the study; this connects to a sociological interpretation of ecology, where human relationships are emphasised from an interpersonal dimension. As previously explained, a PLC brings together a group of like-minded professionals with the aim of learning with and from each other. It is a 'community of practice' (Lave & Wenger, 1991) with learning at its core. A detailed discussion about the CARE PLCs from an ontological angle is given in Hall and Turner (2024); below, a more practical focus is taken. Case studies were constructed by each partner detailing the PLCs' formation, maintenance, and outcomes in each country and then comparisons were drawn between the cases. The research aimed to investigate the opportunities and challenges presented by the PLCs; what social and physical factors foster or hinder communication, positive relationships, teacher confidence, and competencies and perceptions of success in effectively fusing art education with sustainability education.

Each CARE partner country established their own PLC for the teachers during the training programme phase and these involved all research participants within that phase of the project. In Malta there were 25 participants, Cyprus had 24, England had 24, and Greece had 47. All the participants in Malta and Cyprus were fully qualified teachers; in Greece 6 of the 47 were trainee teachers, and in England all 24 were trainee teachers. To exemplify one approach taken, what we referred to as mini-PLCs created with the participants in England, three trainee teachers were grouped with a research team member (a university tutor). This approach was designed to maximise interactions and foster a closer sense of community and co-learning. A similar approach was taken in Cyprus.

Within the mini-PLCs in England, a model of peer learning envisaged the research participants regularly working on activities with a tutor. We set up a CARE programme team in Microsoft Teams and channels for the six mini-PLCs within this. A few participants contributed to this platform, sharing resources they had found, activities they had taught, or asking questions. However, it was tutors who were most active. It was rather disappointing that this aspect of the training programme was not more successful; we thought the participants would enjoy connecting with each other, but online professional communication is not always as popular as in-person exchanges (Hall, 2015), and it was significant that some of those interviewed noted the challenges of using the Teams programme technology.

The mini-PLCs used within the training programme phase of the study and were part of the broader PLC which included all participants and researchers. When the project moved into the phase of implementing the art units in schools, the PLC became smaller (e.g., in England, from 24 participants to 6 participants) but in some ways stronger because member communication increased, and we were able to regularly schedule synchronous meetings which served the purpose of a study group (Hord, 1997). Importantly, the teacher participants were helped to engage in practitioner enquiry (Baumfield et al., 2012), acting not only as a teacher but also as a researcher—again, this is evidence of the formation of a new 'practice architecture' (Kemmis, 2022). For example, the data about children's learning from the art units in schools was collected by the teacher participants, as gatekeepers, who were supported in ethical approval processes and some simple qualitative research methods via additional online training.

According to Hord (1997) the required conditions to form successful PLCs include: supportive and shared leadership, collective creativity, shared values and vision, supportive conditions (physical conditions, people capacities), and shared personal practice. The aim from the start of CARE was for the research teams to foster these conditions, responding to ongoing participant feedback. Despite challenges, not least the restrictions presented by the COVID-19 pandemic, such as the forced move to predominantly online interactions, all partners were able to create and maintain successful PLCs and there was clear evidence of Hord's conditions being achieved (Hall, 2022).

Crucially, all partners found that shared values and vision, such as about the importance of both art education and sustainability education, were key motivations for the CARE participants. We can see an obvious connection here to Bronfenbrenner's (1979) macrosystem of overarching values and beliefs, the synergies within which are crucial to 'make present what causes practitioners to think and feel and act' (Stengers, 2005, p. 195). Regarding people capacities, the PLCs provided emotional as well as practical support. For example, one of the Maltese participants who was interviewed described their PLC as 'a place which I have been longing for … a place where you would see people who are enthusiastic about the work they do, and they would be very committed to their activities' (Hall, 2022, p. 17). In terms of a negative example of people capacities, some of the participants in England described a sense of 'hostility' in placement environments—whilst the PLC provided an antidote to this, it is nonetheless worrying that they felt unsupported in conducting their CARE teaching. To further explain, teachers in placement schools had not selected to be involved in the research, rather it was the trainee teachers' choice. The issue was typically a clash in perceived status of art education and sustainability education—lack of shared values and vision.

Notably, the ecologies within the PLCs reflect some of the teacher competencies required for promoting sustainability education through art education as identified by Pavlou and Kadji (2021): action, attentiveness, creativity, criticality, decisiveness, empathy, futures-thinking, participation, responsibility, systems-awareness, transdisciplinarity, and values. For example, in terms of participation, in Malta, Cyprus, and Greece the PLCs were used as a way to organise the co-creation of art units in schools through shared decision-making. The research participants planned the units together as a formal aspect of the CARE training programme. In the English research team, because we were mindful of workload for our trainee teachers, we took the decision that art units would be planned in a more flexible timeframe outside of the CARE training programme. However, mutual support was still encouraged.

Conclusion

This chapter has reflected on a recent Erasmus+ research project, *Visual art education in new times: Connecting Art with REal life issues* (CARE, 2019–2022), through two main ecological lenses. To begin, an overview of Project CARE was offered and then attention was given to how ecology was interpreted for the purpose of this chapter from both biological and sociological angles. Further, a new 'practice architecture' (Kemmis, 2022) was discussed in terms of the fusion of art education and sustainability education and the notable dual role of the adult participants as teachers and learners. The chapter provided some exemplification of what I chose to refer to as ecologies within the project's knowledge base, followed by a consideration of 'ecologies of practice' (Stengers, 2005) within the CARE PLCs.

Returning to the new 'practice architecture' (Kemmis, 2022), I suggest that the interdisciplinarity of Project CARE was one of its key strengths as its core focus was symbiotic; art education and sustainability education were given equal attention in the research and both equally valued. Further, as an intervention, the CARE research participants were positioned as both teachers and learners, regardless of whether they were trainees or more experienced teachers. University tutors, alike, were positioned as both teachers and learners. According to UNESCO (2023a, para 1), sustainability education 'is a lifelong learning process and an integral part of quality education. It enhances the cognitive, socio-emotional and behavioural dimensions of learning and encompasses learning content and outcomes, pedagogy and the learning environment itself'. This position aligns with how we may choose to describe the value of art education, especially in terms of its vital contribution to holistic learning (Hall, 2023; Pavlou & Kadji, 2021).

Writing this chapter has made apparent that an 'ecological' analysis is potentially a large undertaking, and I cannot state that my discussion here is in any way comprehensive. For instance, I have discussed the ecological aspects of the CARE training programme knowledge base in terms of environmental issues, but I have not discussed the other (connected) aspects of this knowledge base, encompassing economic, social, and cultural issues. However, I have communicated the significance of *some* important ecologies within the CARE research, expanding on previous reported findings. Project CARE was a unique study successfully fusing art education and sustainability education in primary schools in four European countries, and encouragingly its legacy is now being built upon with a digital learning emphasis across other art forms.

Conflict of Interests (1) The author declares there are no conflicts of interest. This study was funded by Erasmus+ (ref: 2019-1-CY01-KA203-058258).

References

Aggarwal, D., Esquivel, N., Hocquet, R., Martin, K., Mungo, C., Nazareth, A., Nikam, J., Odenyo, J., Ravindran, B., Kurinji, L. S., Shawoo, Z., & Yamada, K. (2022). *Charting a youth vision for a just and sustainable future*. Stockholm Environment Institute. https://doi.org/10.51414/sei2022.010

Baumfield, V., Hall, E., & Wall, K. (2012). *Action research in education: Learning through practitioner enquiry*. Sage.

Bronfenbrenner, U. (1979). *The ecology of human development: Experiments by nature and design*. Harvard University Press.

DfE. (2022). *Sustainability and climate change: A strategy for the education and children's services systems (online)*. Retrieved July 23, 2022, from https://www.gov.uk/government/publications/sustainability-and-climate-change-strategy

Fowkes, M., & Fowkes, R. (2022). *Art and climate change*. Thames and Hudson.

Hall, E. (2015, July 7–9). *Creative communities: Professional networks in art and design education*. Paper presented at the 14th International Society of Education through Art European Regional Congress.

Hall, E. (Ed.). (2022). *Professional learning communities in a project connecting primary art education with education for sustainable development*. Report of intellectual output 5, project CARE. Frederick University, Cyprus. http://care.frederick.ac.cy/docs/CARE_IO5.pdf

Hall, E. (2023). A green manifesto for art, craft and design education. *International Journal of Art and Design Education, 42*(4), 611–621.

Hall, E., & Turner, C. (2024). An aesthoecological approach to professional learning communities. In R. Vella & V. Pavlou (Eds.), *Art, sustainability and learning communities: Call to action* (pp. 67–80). Intellect.

Hasio, C., & Crane, T. J. (2014). Teaching art a greener path: Integrating sustainability concepts of interior design curriculum into the art education curriculum. *Art Education, 67*(6), 35–39.

Hickel, J. (2022). *Less is more: How degrowth will save the world.* Penguin.

Hord, S. M. (1997). *Professional learning communities: Communities of continuous inquiry and improvement.* Southwest Educational Development Laboratory.

Inwood, H., & Kennedy, A. (2020). Conceptualising art education as environmental activism in preservice teacher education. *International Journal of Art and Design Education, 39*(3), 585–599.

Ioannidou, M., & Mitakidou, S. (Eds.). (2022). *Children and teachers at work: Facing sustainability challenges through the visual arts.* Frederick University.

Kemmis, S. (2022). *Ecologies of practices: Co-production and interdependence.* Springer Texts in Education.

Kozulin, A. (2023). *The cultural mind: The sociocultural theory of learning.* Cambridge University Press.

Lave, J., & Wenger, E. (1991). *Legitimate peripheral participation in communities of practice. Situated learning: Legitimate peripheral participation.* Cambridge University Press.

Lent, J. (2021). *The web of meaning.* Profile Books.

OED. (n.d.-a). Ecology. In *Oxford English* dictionary.com . Retrieved September 2, 2023, from https://www.oed.com/dictionary/ecology_n?tab=meaning_and_use#5964959

OED. (n.d.-b). Environment. In *Oxford English* dictionary.com. Retrieved September 2, 2023, from https://www.oed.com/dictionary/environment_n?tl=true

Pavlou, V. (Ed.). (2022). *What is really happening now? Connecting art education with real life issues.* Frederick University.

Pavlou, V., & Kadji, C. (2021). Enhancing arts education with education for sustainable development competences: A proposed framework for visual arts education educators. In E. Wagner, C. Svendler Nielson, L. Veloso, A. Suominen, & N. Pachova (Eds.), *Arts, sustainability and education* (pp. 217–235). Springer.

Peacock, A. (2004). *Eco-literacy for primary schools.* Trentham Books.

Stanger, N. (2011). Moving "eco" back into socio-ecological models: A proposal to reorient ecological literacy into human developmental models and school systems. *Human Ecology Review, 18*(2), 167–173.

Stengers, I. (2005). Introductory notes on an ecology of practices. *Cultural Studies Review, 11*(1), 183–196.

Stibbe, A. (2021). *Ecolingusitics: Language, ecology and the stories we live by* (2nd ed.). Routledge.

UNESCO. (2017). *UNESCO moving forward: The 2030 agenda for sustainable Development*. Retrieved July 11, 2022, from https://unesdoc.unesco.org/ark:/48223/pf0000247785/PDF/247785eng.pdf.multi

UNESCO. (2023a). *What is education for sustainable development?* Retrieved April 5, 2023, from https://www.unesco.org/en/education-sustainable-development/need-know

UNESCO. (2023b). *Culture and education*. Retrieved November 13, 2023, from https://www.unesco.org/en/culture-education?hub=66775

Vella, R., Caruana, C., Gatt, I., & Zammit, C. (2022). Part I: The guide. In R. Vella (Ed.), *Integrating principles of education for sustainable development into visual arts education in teacher education programmes* (pp. 9–48). Frederick University.

CHAPTER 8

An Ecology of the In-Between: Liminalities in Practice Research and Material Studio Pedagogies

Ben McDonnell and Molly Pardoe

Introduction

This chapter is about sharing and making space. More specifically, it is about the importance of sharing physical, material space for those involved in practice and research within the context of higher education and the unique possibilities for learning which this space presents. What collaborative practice(s) does the studio allow? What ecologies are these practices situated among? And, what can be learned through noticing the value of the 'in-between' of these ecologies?

We suggest that the shared space of the studio can become a site for novel, relational pedagogies, which we define as being 'in-between'. This space enables participants to negotiate their pedagogical experiences with the institution at large, and find ways to playfully resist the neoliberal logics which are steadily encroaching on higher education institutions. We

B. McDonnell • M. Pardoe (✉)
Educational Studies, Goldsmiths, University of London, London, UK
e-mail: bmcdo001@gold.ac.uk; mpard001@gold.ac.uk

M. Matthews (ed.), *Ecologies in Practice and Learning: Arts Interventions in the Earth Crisis*, Palgrave Studies in Educational Futures, https://doi.org/10.1007/978-3-031-82338-1_8

see neoliberal logics as those which seek to replace the values and practice of co-operation with those of individualism and competition (Ridley, 2017).

In the first section, 'Ecologies of Practice', we map out what the particular ecology is that exists between our practices and within the studio we share. We start from the position of relation through an entanglement of our own practices and this methodological space. We suggest through this that the ecology of in-between is inherently present by putting our practices in relation. We will discuss this methodological environment as a reading of Stengers' ecologies of practice framework (2005). From this positioning, our discussion is enabled through a method of noticing as a process that activates the in-between as space for collaboration. Noticings that occur throughout the text coalesce around two of the larger overarching themes emergent from the practice. The first is the idea of the ecology of the 'in between' which we build upon and iteratively enact throughout the text, as a way of testing/showcasing the boundaries of this space as it stands; the second is that of an ecology of practice that sits within, but operates broadly in resistance to, higher education institutions. Further, through noticing some examples of this modulating in-between ecology, we expand upon the constellation of the in-betweenness as a dynamic, liminal space, which is pedagogical, emergent, and inherently resisting the boundaries it is situated within. From these noticings, we draw conclusions and propositions that stem from explorations of ecology of the in-between.

Ecologies of Practice

We want to conceptualise the studio-making process we have shared as an ecology of practice (Stengers, 2005). Through this lens, Stengers offers a way of articulating practice, not as it is, but rather through the possibilities of what it could be. This enables us to think through our practices in how they generate vitality and meaning beyond the immediate act. Moreover, this conceptual provocation expands the ethos of our practices: Stengers' ecology of practice theory offers tools for thinking 'that address and actualise this power of the situation, that make it a matter of particular concern, in other words, make us think and not recognise' (ibid., p. 185). Understanding the studio space as an ecosystem that we are entangled with is to see it as a place that breathes, one that lives and responds through and beyond our specific encounters, one that has rhythms of its own. It draws attention to the diverse forms of the studio, human and nonhuman, tangible and intangible, as inextricable from the encounters and boundaries we actualise in this chapter. It also emphasises that this space cannot be replicated, nor is it a space that is crystallised. Instead, it is an emergent dynamic, continually in-process, and activated through our mutual investment.

Offering an account of our specific site we work in can provide propositions through which collective, emergent practices are mobilised and recycled for spaces of learning as micro-resistances. This site is simultaneously dependent on our individual approaches but does not necessarily attend to the individual. We are therefore situating our observations or 'noticings' in this chapter within this juxtaposition, which aligns our intention with Stengers. Our propositions here are to observe what our practice ecology is becoming and how it might connect to broader ecosystems of praxis, rather than justify what it is (2005, p. 186). This chapter is, therefore, considered a mapping of our studio ecosystem, with its rhythms of agency as a site for potentiality beyond the specificity of our practice.

Nodes of Relation

We have written collectively about intangible materialities previously (Dixon et al., 2022), where we establish intra-action (Barad, 2007) as a useful term to encapsulate Baradian methodologies more broadly. Intra-action navigates discontinuities and diffractive, enfolding, space-time-mattering theory–material relationships that position our meeting point as an entanglement of parts unfolding together. This fertile, shared ground is where the process of mapping potential relationalities between these particular spatial enactments is situated. Through forming these connections between practices, we understand the larger collaborative space as one that is not directly related to our individual research interests but inhabits an area in-between.

But there is no practice or conversation without the environment (Stengers, 2005). Coming together, emerging from the everyday sensory experience of sharing a space for making (a space *in* the making), is central to our discussion, as all other collective activities of negotiating our presence in the broader university system arise out of this space sharing. Our studio site is situated on the outskirts of our university campus, and the space itself has a rhythm that relates to the subject area we are embedded within. It has high-intensity and low-intensity moments throughout term-time. As PhD Art Practice and Learning candidates, we adopt a long-term marginal positioning within the institution. The studios themselves are shared with the MA Arts and Learning. This meant that our space was never ours, but took different forms, adopting different scenery, different frequencies, different occupations. We nestled into corners, in our margins. This site is reminiscent of similar creative practices taking on an ecological positioning: Dance House Organisms, for example, describe their space as depending on different temporalities, rhythms, and bodies to produce learning encounters (Althammer, 2023). Roberts and Strandvad (2022) think about artist residencies as creative ecologies, also highlighting the processes of cultural production being situated simultaneously within the temporal navigation of the residency environment itself alongside the wider ecosystems of art institutions. Like the studio space, those who enter and exit are both situated within the ecosystem of our shared institution but also bring a shared intention for the nexus between practice and learning, as well as a different temporal rhythm to practice and learning as cultural and pedagogical inquiries.

In *The Production of Space*, Henri Lefebvre categorises social space into *Spatial Practice*, social interactions—*Representation of Space*, that is, the design and planning of space, and *Representational Spaces*, those of art or *images and systems of symbol* (Lefebvre, 2016, p. 39). In the case of our particular ecology of practice, the studio can be thought of not just as a container but part 'spatial practice' and part 'representation of space.' This reimagines the studio as both a designed and planned space which we occupy but also a place made by (not a receptacle of) active, social interaction. In this way, the studio is both a site as a place to congregate and an active, temporal part of our collaborative practices. Susan Orr and Alison Shreeve describe this representation of studio space as a social mix in higher education: 'A [studio] space may not seem like pedagogy, but in its widest sense the studio helps structure what can and does take place when students learn' (Orr & Shreeve, 2017, p. 90); instead, the social space of a studio is where 'students create a social learning environment discussing amongst peers' (ibid.), highlighting the planned representation and social learning that occur with(in) the studio. Orr and Shreve start to describe the studio as pedagogy, a space that encourages and is part of collaborative practice in which all practitioners learn from each other. We follow this position to develop around pedagogy as inherent in the processes of collaborative material relation and dependent on the studio in the ecology of practice outlined here.

Method of Noticing

To understand the emergent, pedagogical space, we adopt a speculative "noticing" method, where drawing out resonances, nodes, connections, pivot points and other important moments in the research is an ongoing, in-process practice. Noticing does not seek to draw out conclusions, and counters the need to articulate knowledge through the notion of resolution; instead, the continuous, collaborative and emergent practice described in our research activity is enacted here. Noticing *draws attention to* rather than attempting to didactically describe and ascribe meaning to a past event.

Learning through noticing involves the learner in constructing their own conceptions of appropriate practice through an understanding of the knowledge domain on which their professional practice is based, on their own observations of practice and on their making sense of these within a social context or community of practice (Rooney & Boud, 2019, p. 443). Rooney and Baud note that observations (or noticings) put into relation with a social context help to define (or make sense of) a practice.

Noticing-as-method is key to understanding the in-between site as an ecology, as it emphasises that this space is in flux and therefore cannot be replicated nor crystallised and retrospectively looked 'at'. Instead, noticing is offered as a way of understanding in-becoming, an emergent dynamic, continually in-process and dependent on our mutual investment. We see these conditions as being attuned to our in-becoming ecology between bodies and materials (Deleuze & Guattari, 1987); therefore, though it cannot be replicated, it can be considered an embodiment of process and as evidence of the capacities for spaces of learning that are foregrounded on a nurturing intra-action (Barad, 2007). It is simultaneously dependent on our individual approaches but necessarily does not attend to the individual. It is the space where collaboration turns into collective practice, and it is distinct but contingent on the sociomaterial dynamics of our own research. Coming together, emerging from the everyday, sensory experience of sharing a space of making (a space *in* the making), is central to our discussion, as all other collective activities of negotiating our presence in the broader university system arise out of this space sharing.

In addition, the method of noticing in relation to the social context of our ecology of practice has facilitated the development of an exploration of everyday resistances—an ecology that operates in resistance to, but is simultaneously dependent on, the higher education institution (HEI). Noticing recognises both the need for and the pull against the hegemonic structures that we work within. It is also a practice where we take responsibility for paying attention as best as we can (Stengers, 2005) in order to discern where our practices extend to, as well as where/what they are limited by. As part of an academic institution and the academy more broadly, PhD students traverse the multiplicitous boundaries between inside and outside relation; in this chapter, this positioning arises in the specific activities we enact, our approaches when taking responsibility for our learning, and our ethical grounding when developing our praxis through writing and being within systems much greater than our little space. It enables us to think through our entanglements from within our environment, where we simultaneously notice how we expand into the space we are given whilst meeting/resisting our encounters with boundaries of this same site. *Noticing as resistance* does not criticise but instead looks for the cracks from which we can nourish and contribute to progressions of our communities. These alternative modes of criticality result from a 'yearning for sustainable futures' (Braidotti, 2019, p. 192); indeed, in noticing our micro-practices of everyday studio life, we are enacting an ethical affirmation of the cracks between the subject area and institutional structures we are constituted by. It is from these cracks that our "in-betweenness" arises. The next section describes the studio crit as a process of the ecology we describe, beginning by placing the reader in the studio through a short piece of qualitative text (Fig. 8.1).

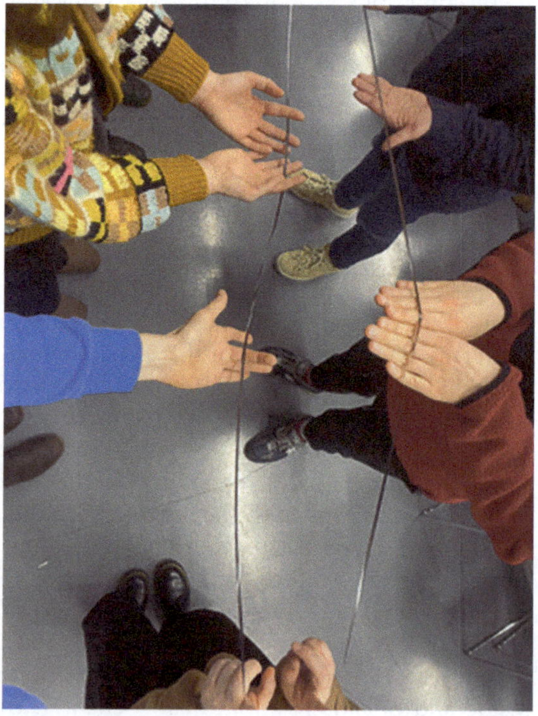

Fig. 8.1 Ecologies of the in-between in studio practice

Studio Crit

Crunch. Then silence. It's another one of those dilated moments, but this one feels different. Cruuuuuunch. We watch as the student lifts her foot, slowly but deliberately. Balanced for a second on one leg, she glances up at her audience with a smile. We don't know what's about to hit us. She shifts her weight and brings her shoe down onto the piece of charcoal. Black shoes, black charcoal, polished grey floor. We're all focused intently on the scene unfolding. There's a pause as the shoe meets the resistance of the charcoal. It's like one of those pressure crushers for testing the strength of materials. The charcoal gives after a moment's resistance, and the crunch rings out. Chhhhhhrrrrrckkkk. Everyone is rapt. There was a before and an after and we were all somewhere before and now we're all in whatever is after.

Through playing non-directed games, sharing space and engaging materially, eventually we started to look at one another's art. This would happen as a natural extension of sharing space and working on our own individual projects in each other's company. These conversations slowly developed into more regular events, and it became expected that if we happened to all be in the studio on the same day, we would take some time to engage with each other's work and offer feedback. After a few sessions, we started to open these up to MA students in our subject area, PhD students in other subjects and eventually this expanded to working with (in-between) PhD students from other art and education-based courses from different universities. We wanted to hold onto the organic, informal feel of our initial conversations, so we used a series of tried-and-tested techniques (silly games and snack breaks) as a way to disrupt the often uncomfortable formalities which can initiate and accompany interaction with art (Blythman et al., 2007).

The above passage shows how the artist's simple yet powerful action created a real punctuation point in the crit, and how our experience of viewing the work was altered as a result. This was the perfect place to experiment with this sort of thing; the playfulness combined with the careful attention of participants meant that we all felt we could try things out. This particular action had a long and productive resonance, both in the discussions and activities within the meeting, as well as beyond the crit gathering. The material enactment altered the way in which we understood how we were occupying space together within the framework of the crit; it also had an effect on each individual's practice work: it became a part of a performance piece as part of a final degree show, and it was transcribed by Ben M in a piece which explores the relationship between space, pedagogy and listening.

Moreover, this moment represents a potent example of how our exchanges within the container of the studio built relational exchanges *between* individual practices. Dialogue at this time was responsive to each other's work, enacting a methodology where we '[read] insights through one another in attending to and responding to the details and specificities of relations of difference and how they matter' (Barad, 2007, p. 71). These considerations attended to a mutual need for work to be seen, considered and made in some cases, reasserting material presence beyond reflective dialogue. Here we were activating each other's practice as a collaborative tool for learning. In this way, the above passage can be seen as an example of the in-between pedagogy: instead of reproducing knowledges through reflective semiotics, we were response-able to being in relation together through material (Barad, 2007). The dialogic arose from the simultaneous material presence, material exchange and making space. The impetus of being seen and heard, as practice-based researchers in the work-in-progress form of our research, allowed us to drop into an ongoing form of pedagogical exchange. The dialogic materially centred exchange of the crit, held in the container of the studio, became a form of research beyond our individual questioning and concerns.

Boundaries: Forms of Resistance

We are still learning through doing within these considerations, but dwelling in-between makes room in the cracks or 'margins of manoeuvrability': by being here more, we are able to make more room (Massumi, 2015, p. 21). Making room through doing-together mobilises the praxis to the boundaries we sit within. Through establishing this cross-level and cross-institutional research community around the material engagement of the crits, we noticed a need for further discursive engagement specifically around practice as research. In addition, PhD students in our department are currently obligated to attend qualitative and quantitative methods courses, but practice researchers were left feeling alienated at the start of their research journey, as there was no structural provision for providing a critical framework around practice research. This feeling of institutional imposition pushed us into structured pedagogic environments that had little to do with the community we were seeking. Noticing a gap in our collective learning became a manoeuvrable force through which we built, proposed and now contribute to as a funded practice-based research methods course offering. These noticings were emergent from the in-between relationality of the crits and acted upon Bulley and Şahin's practice-based research report (2021), which notes the lack of recognition of practice researchers, who often sit at the boundaries of academic institutions, as one of the challenges of this field within the institutional setting.

This in-between relationality is a tangible, institutional result that came from what started as rolling tape on the floor. It shows the potentialities we have before us now: more material–theory–social exchange contributes to our modulations. Expanding into the real-life problematic, our studio practices work beyond, alongside and within the wider ecology. The research methods course put into practice the possibilities of practice-based research as resistance to top-down systems of learning. Staff, students and participants from multiple disciplines shared a table and built understandings of practice, conversationally, as if sitting around a dinner table making notes on the tablecloth. This showcased practice-based work as an 'opportunity [for] dialogue with actual people in actual real places, about actual questions that are actually urgent for them' (ibid., p. 65). These meetings were a purposeful enactment of practice-based research as a site for engaging with the pedagogies of Paolo Freire (1970): we are guided as investigators of potentialities for being 'transforming humans'; sitting and learning through each other (rather than *from* someone as a knowledge centre) in relation to material and theory, we invest in collaborative invention and re-invention of knowledge through a dialogic. This further extends into speculative pedagogies, where 'imaginative and experimental propositions or questions (propositioning) [aim] to construct a future whose efficacy and relevance will have to be negotiated' (Atkinson, 2022, p. 76). Pedagogical practice in this time became synonymous with resistance, as we were mutually entangled by the knowledges we share. This course is indicative of the creative potential of noticing, the pedagogy of the in-between and the fruitful spaces of expanding the network of these ecologies into necessary and emergent contexts (Fig. 8.2).

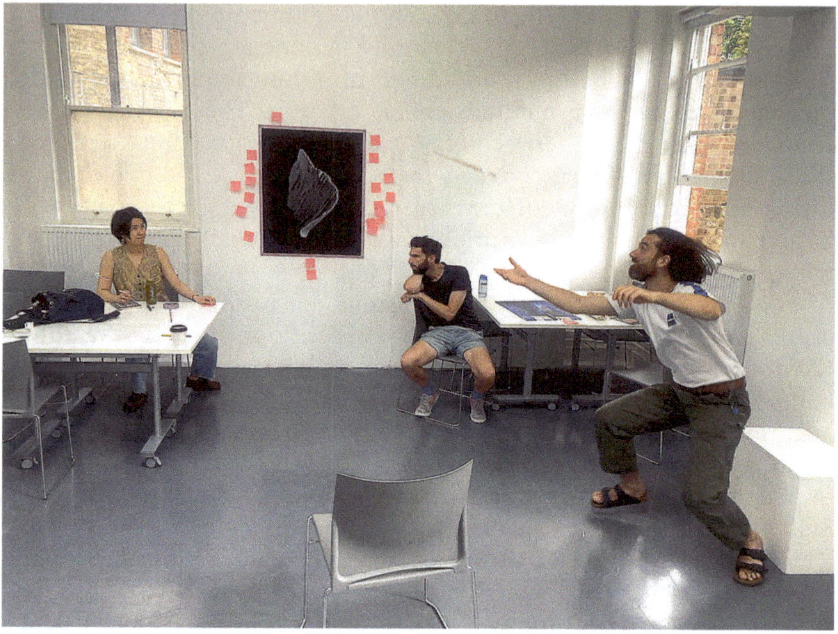

Fig. 8.2 Noticings

Repositioning Competition

Although less tangible, another emergent resistance of this in-between is a site that diffuses competition into collaboration. It uses the nodes of uncertainty, individual practice and forces of 'progress' and repositions them into part of a wider shared ecology, where we work in perspective of as well as rather than an instead of (Stengers, 2005) to motivate research practice. We are engaging in a space that can create moments of pedagogical intrigue that are genuinely creative/novel and that are only possible when there is a mutual encounter with systems that we seek to resist. These encounters are non-hierarchical, experimental, creative and novel. This positioning is subtle but fundamentally divergent from neoliberal mindsets, which cultivate notions of competition to get ahead, be the best, or do more as a practice, all of which assign success to those who have been seen to be the most productive. Academic ranks, silos of courses and departments, a push toward individualism and employability, focused outcomes, etc., of higher education institutions are revealed through the resistances encountered when working collectively within this mechanism.

These systems, however, neglect to notice the impact of how 'doing the most' gets done and leaves little space for in-between moments of resonance, rumination, and unexpected discoveries. These conditions have been driven by the intensity of contemporary university settings, particularly in arts-based disciplines (Manning & Massumi, 2014).

Our in-between ecology offers an alternative in the way it relates to this environment, thereby showing us otherwise through pedagogical relation to the boundaries of our context, as well as dimensions of change and resistance from within this site (Hickey-Moody & Page, 2015). By aligning to the ecologies of our shared environment, we participate within these systems but from a different condition to 'how' we make our way through. As PhD students, we train individually to research; noticing that we can also make space together pedagogically during our training builds upon a sustainable ethics for,

> non-unitary subjects, [which] rest on an enlarged sense of inter-connection between self and others, including human or 'earth' others, by removing the obstacle of self-centred individualism on the one hand and the barriers of negativity on the other. (Braidotti, 2019, p. 190)

This is a long-term orientation that diverges from establishing our research practice alone, to establishing our research amid our shared space. In the praxis we have outlined here, we see ideas/outputs as constantly in-becoming and in-relation, which enables generosity: we are not withholding, hoarding or taking ownership of our research environment. Instead, we share our own practice *as well as* build upon the ways these intra-act with each other, knowing that in-becoming is striving in a processual relation to knowledge, rather than crystallising discovery as something to hold onto. We are affirming space for the "*as well as*" in this ecology. This practice 'entails the decision never to accept capitalist destruction as freeing the ground for anything but capitalism itself' (Stengers, 2005, p. 185) and instead grows within alternatives that think and act pedagogically for the future together (Atkinson, 2021).

Sustainable Resistances

There is an important limitation to the shapes of this 'in-betweenness' that will be discussed here through an affirmative ethics of hope. This is done by noticing the space, both literal and temporal, that has been afforded to us to enable this work to happen together. Within the framework of in-between ecologies, space is both a commodity and a resource: the environment that we are based within was fundamental to plant conceptual/material seeds within, to allow for play and spontaneity and to put our material in relation with each other without criteria. The annual turnaround with a new group of MA students saw more individuals needing their individually allocated studio spaces this year than in previous years, which meant our physical margins were needed for these occupations. We therefore no longer share space. Temporal conditions of institutional environments are such that our longer-term and creative positionings are pushed further out to the margins, where our presence is one of transient, impermanent occupation (e.g. Manning & Massumi, 2014). Without the environmental grounding and the resource of space, these practices cease to work in a fluid and mobilising way. These contingencies highlight at once the pressing need and lack of space that contemporary universities have available. Allocated departmental and individual processes demarcate zones of practice; such possessive frameworks limit those who are between internal–external conditions. We suggest that these observations prompt another in-between consideration: what would the conditions of an environment made for our ecologies look like? How can we develop institutional in-between sites, where we acknowledge the ways in which 'both the universality of formal knowledge and local processes of experiencing [are] intermingled' (Atkinson, 2022, p. 72) in a way that encourages unexpected and emergent relations rather than routing towards structures that sit atop a mountain of established systems? Adapting institutions to make space for the in-between margins to mingle more often with structural processes might be the very thing needed to enact an alternative to the neoliberal frameworks we so often critique.

Furthermore, these observations bring us further into noticing our ecology and the conditions of our practices together; as space is neither a given nor an expectation, how can we sustain these ecologies in practice without falling to whim and circumstance? What kinds of adaptations allow us to continue an in-becoming force that develops and extends our praxis? How do temporary encounters reframe how we work together? And, lastly, how do we engage with our ethical praxis in developments that build upon future perspectives? We are propagating this in-between practice by using these evolving margins to experiment further, thereby bringing hope into our praxis too (Massumi, 2015).

Propositions—Conclusion

In this chapter, we have discussed collaborative activities of group crits and designing a course that have been enabled by an in-between ecology of collaborative practice. We do not seek to have answers but rather engage with a fruitful exploration of what it means to exist in a materially centred collaboration. Noticing in this way has allowed us to identify the areas of this collaboration that both initiate and grow our modes of working together beyond the practices that we bring into the in-between. The in-betweenness for us is multiplicitous. It is spatial: we work between studios, subject areas and institutions; it is material: we work between our own material practices; it is interdisciplinary: we meet between pedagogy and art. We suggest that it is neither any one of these things, nor is it an intersection of these things; rather, it is beyond and through these nodes of relation that we build our in-between.

Offering noticings on the ecology we created within the university and the potential benefits to this approach in terms of providing alternative ways of working, thinking and doing can subvert the neoliberal logics embedded in public institutions. Therefore, noticing the in-between and working into it could be articulated as a resistance against these neoliberal logics, which tend towards arboreal structures, productivity over play, and individualisation over collectivity. We have created these ecological pockets of learning together, which could be a model for how engagement together can resemble knowledges that go beyond the institutional frameworks we are situated within.

Having started to notice the in-between and its importance in working collaboratively, it became increasingly evident that inhabiting this in-between together is a critical position, as it presents an alternative way of being together in practice. Through making an in-between space, it is necessary to be in resistance to that which it is expanding against, within and among. Conversely, it is not possible to be in resistance without the structures it is resistant to, which presents an interesting 'in-between' tension inherent to this positioning. This acknowledgement offers a framework for future practice.

Our work evidences the ways in which this in-betweenness is located beyond the multiple nodes that we are positioned within, forming a collective practice that breathes and expands into/beyond the structures in which we work. We suggest that the way of working that we have presented here contributes a model for building sustainable, critical communities which are able to resist institutionalisation through playful experimentation.

Conflict of Interests (4) The authors have no conflicts of interest to declare that are relevant to the content of this chapter.

References

Althammer, M. (2023). Dance house organisms: Reflections on institutional ecologies (for a future dance house inEurope). In N. Gareis & N. Haitzinger (Eds.), *Curating dance: Decolonizing dance* (Issue 55). OnCurating. https://www.on-curating.org/issue-55-reader/dance-houseorganisms-reflections-on-institutional-ecologies-for-a-future-dance-house-in-europe

Atkinson, D. (2021). Inheritance, disobedience and speculation in pedagogic practice. *Discourse: Studies in the Cultural Politics of Education, 43*(5), 749–765. https://doi.org/10.1080/01596306.2021.1975911

Atkinson, D. (2022). *Pedagogies of taking care: Art, pedagogy and the gift of otherness.* Bloomsbury Publishing.

Barad, K. (2007). *Meeting the universe halfway: Quantum physics and the entanglement of matter and meaning.* Duke University Press.

Blythman, M., Orr, S., & Blair, B. (2007). *Critiquing the crit. Project report.* Higher Education Academy. Retrieved August 12, 2024, from https://www.academia.edu/586074/Critiquing_the_Crit

Braidotti, R. (2019). *Posthuman knowledge.* Polity Press.

Bulley, J., & sahin, Ö. (2021). Practice research - Report 1: What is practice research? and Report 2: How can practice research be shared? *Practice Research Advisory Group UK (PRAG-UK).* Retrieved August 12, 2024, from https://ukrio.org/wp-content/uploads/2022-02-09-Bulley-Sahin-Practice-research-reports-UKRIO-Webinar-public-share.pdf

Deleuze, G., & Guattari, F. (1987). *A thousand plateaus: Capitalism and schizophrenia.* Trans. B. Massumi. Minneapolis University Press.

Dixon, B., McDonnell, B., & Pardoe, M. (2022). Intangible materialities: Exploring the unseen in practice research. *Holt Journal for Artistic Research, Issue 1 Methodologies.* https://holtjournal.co.uk/Ben-Dixon-Ben-McDonnell-and-Molly-Pardoe

Freire, P. (1970). *Pedagogy of the oppressed.* Seabury Press.

Hickey-Moody, A., & Page, T. (2015). *Arts, pedagogy and cultural resistance: New materialisms.* Rowman and Littlefield International.

Lefebvre, H. (2016). *Rhythmanalysis, space, time and everyday life.* Bloomsbury.

Manning, E., & Massumi, B. (2014). For thought in the act. In *Thought in the act: Passages in the ecology of experience* (pp. 83–134). University of Minnesota Press.

Massumi, B. (2015). *Politics of affect*. Polity Press.

Orr, S., & Shreeve, A. (2017). *Art and design pedagogy in higher education: Knowledge, values and ambiguity in the creative curriculum*. Taylor & Francis Group.

Ridley, D. (2017). Institutionalising critical pedagogy: Lessons from against and beyond the neo-liberal university. *Power and Education, 9*(1), 65–81.

Roberts, K. S., & Strandvad, S. M. (2022). Artist residencies as creative ecologies: Proposing a new framework for twenty-first-century cultural production. In L. McCormick (Ed.), *The cultural sociology of art and music: New directions and new discoveries* (pp. 43–69). Palgrave Macmillan.

Rooney, D., & Boud, D. (2019). Toward a pedagogy for professional noticing: Learning through observation. *Vocations and Learning, 12*, 441–457.

Stengers, I. (2005). Introductory notes on an ecology of practices. *Cultural Studies Review, 11*(1), 183–196.

PART III

Eco-pedagogies in Site-Specific Practice and Collaboration

This section of the book draws attention to arts pedagogies brought into school and university environments in a specific school or stage of learning. Projects with specific schools give a more granular approach to ecologies in practice. There are also theorised playful pedagogies that emphasise posthuman interconnection of the worlds of the human and non-human.

Henrietta Patience argues for physical spaces to be specifically dedicated to biophilic arts practices in secondary schools. Patience discusses a site-specific project developed in enrichment sessions outside the formal art and design curriculum, in a school for 11–18 year olds in London, UK. The 'liveable artworks' created by students are planned to become future ecosystems in themselves. Patience positions a caring and compassionate approach to degrowth and 'slow hope' working at 'micro-scale' to re-harmonise nature. This chapter also, however, discusses how school students need to become more aware of the work of internationally known artists and global climate initiatives, with a commitment to action and engagement.

Terri Newman theorises and practices ecologies of collaboration in post-sixteen education for students aged 16–19. Newman situates the close observation of pedagogies in ecologies of practice that work with the posthumanist dynamics of the classroom as a rhizomatic 'criss-crossed cartography' that can buck the traditions of art and design practice. In this newly becoming world, Newman proposes art and design pedagogies that accept and embrace entangled futures, in which young people are testing out their own practice-based rules for environmental and emotionally aware learning intra-connections.

Kevin Campbell Davidson integrally relates to the Earth Crisis in his teaching and arts practice. In this chapter Davidson argues for the application of 'threshold concepts', with a creative de-centring of human subjectivity, that, for example, recognise communication between human and other-than-human life forms, and interdependent ways of knowing. Davidson discusses how applying these posthumanist concepts, in combination with 'systems thinking games', as cross-curricular arts methods for first year undergraduate students, can assist students to deal with the 'super complexity' of the crisis.

The pedagogies and projects in discussion here explore the precarity of trying to bring in new challenging ecologies in practice within institutional environments. Authors offer particular insights and encouragement to arts practitioners who are navigating and pushing the boundaries of formal educational processes.

CHAPTER 9

A Space for Ecology: An Exploration of the Role Art and Design Education Can Play in Supporting Environmental and Ecological Issues

Henrietta Patience

Introduction—Education's Role

Whilst reflecting on working with students to develop a site-specific artwork with a focus on ecology this chapter puts forward an argument for there to be space within schools and school curriculum for students to develop knowledge, skills and agency to navigate their future, make informed decisions and contribute to sustainability. Specifically, this chapter argues for the power of art and design to deepen engagement, develop critical thinking skills and support interdisciplinary approaches within schools. A case will be made for the need for time and space to slow down and focus on positive and hopeful narratives in problem solving and finding solutions, even when these are on a micro scale.

School students are significantly impacted by the climate crisis, and in some cases climate anxiety, facing uncertainties about the future of the

H. Patience (✉)
Educational Studies, Goldsmiths, University of London, London, UK
e-mail: h.patience@gold.ac.uk

M. Matthews (ed.), *Ecologies in Practice and Learning: Arts Interventions in the Earth Crisis*, Palgrave Studies in Educational Futures, https://doi.org/10.1007/978-3-031-82338-1_9

planet and their roles in addressing environmental change. A recent study by Pearsons highlighted that teachers have seen a rise in students' awareness and anxiety over global issues such as world conflict and climate change, and that 61% of teachers felt that the current education system is not successfully developing tolerant, sustainably minded global citizens of the future (Pearsons, 2022). This is echoed by the work of Teach the Future (2023), a campaign for climate education to permeate the entire curriculum, composed of students from all four nations of the UK and supported by the student-led charity SOS-UK (Students Organising for Sustainability UK). They believe that students should be taught about the climate emergency and ecological crises: how they are caused, what we can do to mitigate them and what our future lives and jobs are going to look like because of them.

The Climate Change Conference, Conference of the Parties to the United Nations Framework Convention on Climate Change (COP26), took place in Glasgow in 2021. The UK Department for Education (DfE) led the conversation on how to fight climate change through education and acknowledged that they have an important role to play in all aspects of sustainability (DfE, 2022). In the recently published Department for Education policy paper 'Sustainability and Climate change: a strategy for the education and children's services systems' (DfE, 2022), there is a focus on preparing young people with green skills for green jobs and the green economy that is anticipated to touch every career. However, with the National Curriculum's limited reference to climate change (National Curriculum, 2014) and its absence from the Office for Standards in Education, Children's Services and Skills (Ofsted) new education inspection framework (Ofsted, 2022), there is clearly a long way to go. In both Scotland and Wales, the curriculum includes more explicit references to climate change. In Scotland, there is a Learning for Sustainability (LfS) framework that ensures that all learners can develop knowledge and skills related to sustainability and climate change, which are integrated across subjects (Scottish Government, 2023). The curriculum in Wales aims to develop ethically informed citizens who are knowledgeable about their impact on the world (Curriculum for Wales, 2022).

One of the great liberating strengths of art and design as a school subject and art in a wider social context is that it can take its subject and inspiration from the almost endless sum of human experience. As 'Teach for the Future' points out, much of a school's existing curriculum content can be edited, reframed and adapted (Catallo et al., 2022); this is never more so than in art and design where a broad curriculum can provide a platform for challenging important political issues, including social justice, equity, environment, sustainability, citizenship and diversity (Downing & Watson, 2004; Burgess, 2003; Gadsby & Bullivant, 2010; Trowell, 2010; Addison & Burgess, 2021). Exploring themes of ecology and sustainability is a focus and priority in my own work with school students. More explicit references to climate change can be found in both the Scottish and Welsh national curricula.

Changing Ecology/New Roots

Having been invited to run a series of art enrichment workshops, I first entered the gates of a shiny new school in 2021. The school's founder members were about to start their GCSEs, and the workshops would support students further studies in their GCSE portfolio (AQA, 2022). This school is on the site of a former 314-year-old school, controversially closed amid claims of falling pupil numbers and a rising financial deficit. What was once a thriving school community was disbanded and students displaced, having to find new schools throughout the borough. The site remained abandoned and derelict for the next five years, though amid the destruction, nature reclaimed the space, rewilding to reintroduce wildlife.

In 2018, The Education and Skills Funding Agency (ESFA) approved plans for a new school to be built on the site to ensure enough future school places. In 2019, work started on demolishing the original structures and building a new school. The environment, climate and culture are essential for a school and its students to thrive. Parallels can be drawn between ecosystems found in nature when we think about how those in schools work. Bronfenbrenner's Ecological Systems Theory of Human Development (1979) examines the complex interactions and relationships between an individual and their multiple social and physical surroundings during adolescent development. The immediate environment in which an individual operates, in this case, a school, is known as the microsystem (Bronfenbrenner & Morris, 1998). As with all 'natural' habitats in which human beings live and respond, the school is an inescapable part of the environment of all individuals and for some the most formative years of their life (Eggleston, 2012).

I am impressed by the consideration given to sustainable design, materials, ergonomics and aesthetics in recent studies which have shown that students' performance is enhanced in schools with better physical learning environments (Barrett et al., 2019). Biophilic design considers the relationships between nature, science, and the built environment, relating back to the desire for a (re)connection with nature and natural systems. Most recently, biophilic design is being championed as a complementary strategy for addressing workplace stress, student performance, patient recovery, community cohesiveness and other familiar challenges to health and overall well-being (Browning et al., 2014).

The Head Teacher has a strong vision for the school and its students, there is a feeling of optimism as the school, its students and the natural environment grows. If we consider a school as an ecosystem where students, staff and visitors interact with their physical environment, this has real implications for how we organise a school and conduct ourselves within it (Elias, 2016). The main 3-storey block has a figure of eight-like structure, being built around two internal courtyards or atria, each floor houses a different school college, like small schools within the whole, these communities promote a sense of belonging (Osterman, 2000; Pittman & Richmond, 2007).

There is also a vision for the school grounds. Bees, vegetable plots and the saplings that diffuse the area are early signs of the anticipated permaculture (Mollison & Holmgren, 1978). It seemed timely to focus on ecological and environmental concerns, and I proposed that students work collaboratively to produce a site-specific artwork for the school grounds (Fig. 9.1). The designated site is a space for all students, staff, and visitors to engage with the natural world and explore wider environmental issues of ecology and climate change. The brief for students was to design a living artwork that would promote ecology, attracting interaction from insects and wildlife.

Slow Hope

The enrichment sessions ran across two terms. In many cases, students worked collaboratively in pairs or small groups on larger designs though some chose to work on individual pieces. In early sessions, students were encouraged to explore their concerns and share ideas through the creation of disruptive collage to challenge complacency, provoke questioning and conversations. The immediacy and tactility of collage lends itself to less inhibited responses that communicate impactful messages through image and text. The students in these sessions have a wide range of enthusiasms and anxieties; for most, the plight of their planet through climate change

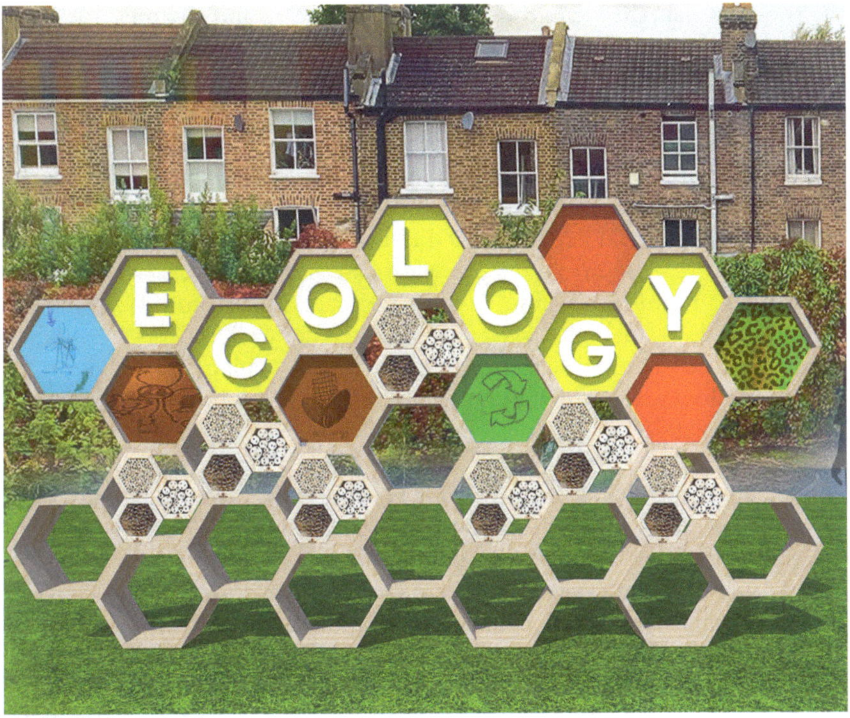

Fig. 9.1 Visualising the artwork

is a priority. The American Psychological Association has described 'a chronic fear of environmental doom' (Clayton et al., 2017). This is a pressure that adds to worries about relationships, self-image, status and all the other adolescent anxieties.

The art department could be a space within which students can express their concerns and priorities through visual communication. Many of these concerns can seem trivial from an adult perspective, or their solutions misguided, but if art educators are to embrace the idea of student agency, they should be part of the mix. It is important that students know that we care about their concerns and take them seriously; we should also remember that hope and imagining a better future are critical. Students will be all too aware that their voice cannot make changes to the problems of the planet, but in their art projects they can and should be responsible for the materials they use, the materials they waste and the messages they are sending.

With the bureaucratic and frenetic nature of schools, the ever-growing pressures on academic achievement, and for teachers to fulfil curriculum requirements, there is little time to develop a pedagogy of care (Atkinson, 2022). The freedom from the parameters of a timetabled lesson lends itself to a collaborative pedagogical approach and a time to slow down, contemplate and connect. This is a different kind of ecosystem, one where students can be present and have the head and physical space to cross-pollinate and find new ways of doing and thinking. Considered here are the principals of Geir Berthelsen's 'Slow movement' and his creation of the think tank 'The World Institute of Slowness', which offers a vision for an entire 'slow planet' and a need to recover reflection, and togetherness (Berthelsen, 1999).

The 'Degrowth movement', a term coined by French philosopher André Gorz in 1972, is also relevant to slow hope. Degrowth questions the compatibility of capitalism with Earth's balance (Duverger, 2020). There is a shared focus on re-evaluating our pace of life. Berthelsen emphasises embracing a slower, more deliberate approach, resonating with the degrowth philosophy that challenges the perpetual pursuit of economic growth. Both philosophies advocate for a shift towards sustainable and mindful living, encouraging critical thinking about societal norms and fostering a sense of balance between personal growth and environmental responsibility. Art and education can be platforms for students to question consumerism, promote sustainability, and reflect on the impact of their choices. The problem for us now as artists, designers and educators is to make outcomes that are sustainable, cheap and desirable, including the awareness that none of our actions are without consequences for the wider world. The micro is indeed the macro when it comes to the environment.

The notion of 'slow hope' (Mauch, 2019) resonates with healthy ecology and sustainable practice. Inspiration can also be taken from the exhibition 'Dear Earth' at London's Hayward Gallery (2023), in which there was a focus on 'resilience', 'care' and 'tending' (Fowkes et al., 2023). The exhibition brought together a collection of works by artists responding to the ecological crisis but also envisioning a better future 'The most impactful art goes beyond outraged protest or a cry for immediate action, and instead deepens our engagement with the subject in ways that ultimately nurture both our understanding and our capacity to act' (Rugoff, 2023, p. 12). As well as art informing us, it has a larger task in the climate crisis and that is to make us the people we need to be to respond in the ways we must (Solnit, 2023). It has the power to evoke emotions, raise awareness and prompt action. Art can be formative and contribute to the development of more confident individuals, fostering a sense of empowerment in addressing complex issues. It has the capacity to shape perspectives, provide knowledge and inspire transformative imaginaries.

In addition to 'slowing down' in the classroom, an important aspect of these sessions was to explore and interact with the outdoor environment for students to escape the school building and reconnect with nature, observing and surveying how ecology interconnects and thrives. Incorporating such opportunities for exploring nature can also improve students' mental and physical well-being, which is crucial for effective learning. This is even more pertinent when research suggests that young people's connection with nature declines in their teens (The United Nations Children's Fund – UNICEF, 2015).

Biomimicry, which involves learning from nature's design and applying those principles to solve human challenges, offers a valuable framework for education. It emphasises sustainability and innovation. The key principles include emulating nature's patterns, using local resources efficiently and fostering systems that support life (Biomimicry Institute, 2024). Working in small groups, students took stock of the wildlife, insects and nature found in the proposed site for their artworks, gaining insight into how their artwork might support the local ecosystem. They used photography to record colours, textures, and surfaces; they collected natural materials to create land art inspired by the work of Richard Long (2023), Andy Goldsworthy (2022), and the Boyle Family (2024), who are best known for their earth studies. Three-dimensional casts of the earth's surface record and document random sites with great accuracy. These works combine real materials from the site (stones, dust, twigs etc.) with paint and resins. The students returned to the classroom, where they pressed and imprinted the materials they had collected on site into clay moulds, forming plaster reliefs to make recordings of their site (Fig. 9.2) while thinking critically about how artists can interact with their environment and draw on nature.

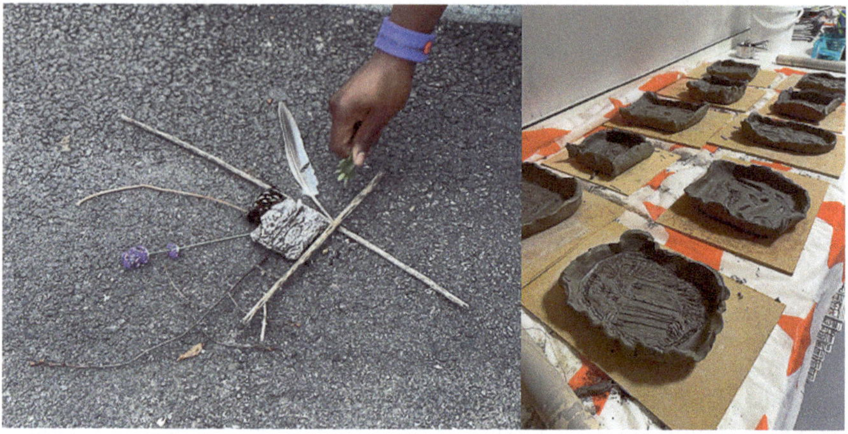

Fig. 9.2 Students creating land art and plaster reliefs

Critical Thinkers

Whilst as teachers we have a responsibility to debate, and to some extent direct thought, this should be to expand not contract our students' thinking. The current Earth Crisis has led some artists and designers to use sustainable materials and create work which raises awareness and expresses concerns. Ecology is a popular theme for gallery programming and art exhibitions—such as 'Our Time on Earth', a major exhibition at the Barbican Art Gallery, London in 2022 (Barbican 2024), celebrating the power of global creativity to transform the conversation around the climate emergency. Most recently, 'Dear Earth' brought together the work of artists from all over the world with a focus on inviting the audience to reimagine responses to the climate crisis (Rugoff, 2023). I invited students to think critically and explore the power of art in communicating important social and political messages.

We discussed the work of contemporary artists, including Olafur Eliasson, and in particular 'The Weather Project' which filled the Turbine Hall at Tate in 2003 (2023). Students were intrigued by Eliasson's use of materials which included humidifiers to create a fine mist in the air via a mixture of sugar and water, as well as a semi-circular disc made up of hundreds of monochromatic lamps which radiated yellow light. We discussed how audiences might respond to this work. The ceiling of the hall was covered with a vast mirror in which visitors could see themselves as tiny shadows against a mass of orange light.

Students were introduced to work that makes political points, such as Ai Weiwei's use of recycled materials for 'Straight' 2008–2012, in which 150 tonnes of scrap metal were salvaged from a school destroyed by an earthquake in Sichuan. This work was designed to highlight the poor quality of government building materials. 'Souvenir from Shanghai' 2012, a cube of concrete and brick rubble from the artist's studio demolished by the government, was set within an antique carved door frame. Environmental activism was also explored through Jenny Kendler's 40-foot-long outdoor sculpture, which displays a hundred reflective birds' eyes mounted on aluminium. The eyes belong to birds threatened by climate change; this artwork was originally exhibited at the Eden Project and more recently shown at the Hayward Gallery.

We also visited Goldsmiths Centre for Contemporary Art's exhibition 'Testament' (Cripps, 2022), where students experienced contemporary art at first hand, a first for many of them. The exhibition filled the entire gallery and featured proposals by 47 artists. Students were invited to consider the artists' proposals and think about what is at stake in tearing down and erecting monuments, and what it might mean to rethink the idea of the monument. Students considered the messages that they would want to communicate to future generations; these included gender equality, cultural diversity, sustainability and protecting endangered species. They developed their ideas further through a workshop that encouraged them to consider monuments of today and throughout history. Ideas were taken back to the enrichment sessions, where students worked collaboratively to produce their own monument maquettes drawing on their own climate themes and ideas of legacy. The development of these ideas also presented opportunities to draw on knowledge from other areas of the curriculum, for example, science, history, geography, and design and technology.

Cross-Pollination

Some of the most impactful ecological art that students were introduced to throughout these sessions brought disciplines together, such as science and art, through the work of Olafur Eliasson (The Weather Project). An art project with a focus on ecology provides opportunities to work collaboratively across subjects, sharing valuable knowledge. In Donna Haraway's 'Staying with the Trouble, she explores the intersections of art, science and activism, highlighting the importance of collaborative approaches to addressing environmental and social challenges. The Crochet Coral Reef Project started in Los Angeles and now worldwide invites participants to create crochet coral reefs to raise awareness of the coral reef ecosystems and the impact of climate change on marine life. The crochet reefs serve as both artistic creations and educational tools, highlighting the beauty and diversity of coral reefs while also addressing the urgent need for environmental conservation. Interdisciplinary projects can encourage students to engage with real-life issues in meaningful ways and recognise the connections between different disciplines and the importance of taking a holistic approach to problem solving, 'staying with the trouble' and working together to navigate complex challenges (Haraway, 2016).

Young people's understanding of the world comes through their ability to make connections; real life is interdisciplinary and complex (Trowell, 2010; Gadsby & Bullivant, 2010; Barnes, 2015; Tanya & Twyman, 2018). However, as discussed earlier in this chapter, there is growing pressure on time in schools which can too often inhibit opportunities for cross-curricular approaches that would allow for innovative exploration of sustainable, environmental and ecological themes. At this school, there is a focus on Science, Technology Engineering, Arts and Mathematics (STEAM). As students developed design ideas for their liveable artworks, it was important to understand more about the proposed site's ecology and how their artwork would attract insects and wildlife to interact with them. We invited the school's STEAM Ambassadors (students from years 7 to 11, who volunteer to regularly get involved in STEAM initiatives competitions and cross-curricular events), who researched the local ecology and gave students advice on insects and wildlife the artworks might attract. Conversations continued throughout the making process, also using Google classroom, students posted their questions such as "Do ceramic glazes contain any properties that might be harmful?" or "What sorts of insects might be likely to interact with these artworks?" along with the ambassadors' research and feedback.

Students interacted with multiple layers of their educational ecosystem, building new relationships across the school years and subjects as they exchanged knowledge. Fostering these relationships resonates with Bronfenbrenner's ecological systems, contributing to a more inclusive and supportive environment, reflecting the macrosystem's cultural and societal influences. It is critical for students to understand that we do not function independently but are interconnected and can influence one another, as with the mesosystem, where interactions between different environmental systems can have a significant impact on individual development (Bronfenbrenner & Morris, 1998). Bronfenbrenner's ecological systems theory emphasises the influence of various environmental systems on individual development, ranging from the immediate microsystem (school and family) to the broader macrosystem (cultural values, societal norms). Posthumanist thinking challenges traditional human-centric views and considers entities as active participants in ecological systems. In this context, posthumanism could expand Bronfenbrenner's theory by including non-human elements in the ecosystem, for example, the impact of technology, animals, plants and even non-living entities on human development and behaviour. Related to this expansion, Haraway discusses the idea of the 'fuller scene', a way of understanding the world that rejects fixed boundaries and categories. Instead of viewing things in isolation, she suggests that we recognise the interconnectedness of all entities, both human and non-human. It invites us to consider the intricate web of relationships and influences that shape an individual and collective existence, highlighting the importance of interconnectedness and collaboration (Haraway, 2016).

Enrichment sessions and digital classrooms present new ways for schools to be able to support these valuable interdisciplinary collaborations. There is potential to make ecology interconnected and relevant in schools, preparing students for green careers. Since the issues of climate change and sustainability are caused by and will affect all areas of our lives, it makes sense to cross-pollinate our education, providing a holistic understanding of the implications.

A Work in Progress

Over the weeks that followed, students used their designs to develop liveable artworks in terracotta, chosen as it is 100% natural, recyclable, long-lasting and weatherproof. Inspiration was taken from the 3D printed terracotta tiles placed along Hong Kong's coastline, designed to revive marine life (Keegan, 2020). Some students chose to add colour to their designs through use of underglaze, which is non-toxic and cheap, as only a single firing is needed, whereas overglaze decoration requires a second firing, albeit at a lower temperature. The enrichment workshops become a space for slow thinking, slow collaboration, slow discussion, slow looking, slow art and ultimately slow hope.

As students walked through the school gates in September, they encountered the new additions to their school grounds, that not only improved their environment aesthetically but also ecologically (Fig. 9.3). A discussion point, a space for contemplation and a reassurance that change is possible were created. What they built and what will happen to it in the future could be archived every season to record the natural changes. It may become more beautiful or more derelict as creatures inhabit the artwork, and it becomes overgrown with plant life and fungi. The outcome is not predetermined but full of hope and possibility. The liveable artworks could become a managed ecosystem, in which future students take responsibility for the interventions they make. There is the possibility of forming a Horticulture Club to oversee this ecosystem.

In relation to the whole planet's problems, this work might seem too small to be important, but it has allowed students the space to think and create in an environmentally aware manner. It has also encouraged collaboration with disciplines other than art for the enrichment of all. Over time, students at this school can appreciate the development of a healthy and thriving environment for which they are temporary stewards, a legacy they will pass on to future generations. Furthermore, the project has encouraged a sense of the students 'taking control' of their environment and being able to make a difference to the school space, which in turn could influence their future decisions, as they come into more positions of responsibility as adults. It is a function and by-product of the work that goes on in school to spread into the home and the wider community. This is especially the case when ecological arts projects involve ideas that are being discussed in the wider world, such as international climate change conferences, and therefore become particularly relatable experiences. In this chapter, I have shown how school projects can be a great 'conversation-starter' for positive change.

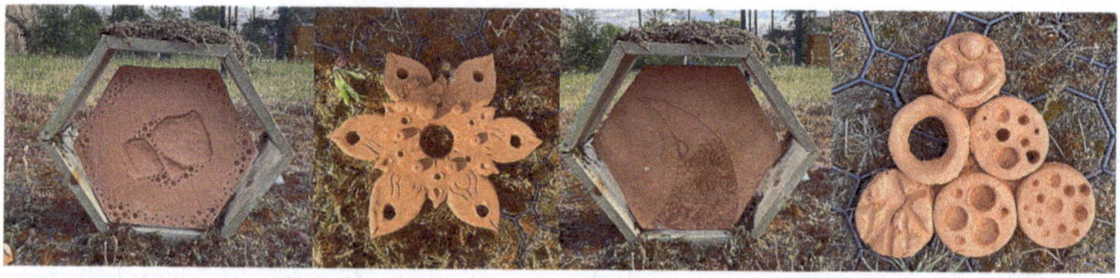

Fig. 9.3 Student artwork on site

Conflict of Interests (4) The author has no conflicts of interest to declare that are relevant to the content of this chapter.

References

Addison, N., & Burgess, L. (Eds.). (2021). *Debates in art and design education* (2nd ed.). Routledge.

Andy Goldsworthy. (2022). Retrieved October 30, 2022, from https://artuk.org/learn/learning-resources/andy-goldsworthy-and-land-art

AQA. (2022). *GCSE art and design specifications.* Retrieved October 30, 2022, from https://www.aqa.org.uk/subjects/art-and-design/gcse/art-and-design-8201-8206/specification-at-a-glance

Atkinson, D. (2022). *Pedagogies of taking care.* Bloomsbury Publishing.

Barnes, J. (2015). *Cross-curricular learning* (pp. 3–14). Sage.

Barrett, P., Treves, A., Shmis, T., Ambasz, D., & Ustinova, M. (2019). *The impact of school infrastructure on learning: A synthesis of the evidence.* International Development in Focus. World Bank. Retrieved August 12, 2024, from https://files.eric.ed.gov/fulltext/ED604388.pdf

Berthelsen, G. (1999). *The world institute of slowness.* Retrieved February 14, 2024, from https://www.theworldinstituteofslowness.com/our-founder-geir-berthelsen/

Biomimicry Institute. (2024). Retrieved October 6, 2024, from https://biomimicry.org/ourmission/

Boyle Family. (2024). Retrieved February 16, 2024, from https://www.boyle-family.co.uk

Bronfenbrenner, U., & Morris, P. (1998). The ecology of developmental process. In W. Damon & R. M. Lerner (Eds.), *Handbook of child psychology* (Vol. 1, pp. 993–1028). Wiley.

Browning, W. D., Ryan, C. O., & Clancy, J. O. (2014). *14 patterns of biophilic design.* Terrapin Bright Green llc.

Catallo, A., Lee, E., & Vare, P. (2022). *Curriculum for a changing climate: A track changes review of the national curriculum for England.* Retrieved August 12, 2024, from https://assets-global.website-files.com/5f8805cef8a604de754618bb/637d2a1bede15e6467a233dd_Curriculum%20for%20a%20Changing%20Climate_%20a%20track%20changes%20review%20of%20the%20national%20curriculum%20for%20England.pdf

Clayton, S., Manning, C. M., Krygsman, K., & Speiser, M. (2017). *Mental health and our changing climate: Impacts, implications, and guidance.* American Psychological Association & Eco America.

Cripps, P. (2022). Testament,. Goldsmiths Centre for Contemporary Art.

Dear Earth. (2023). *Hayward gallery.* Retrieved August 6, 2024, from https://www.southbankcentre.co.uk/whats-on/art-exhibitions/dear-earth

DfE (Department for Education). (2014). *The national curriculum in England: key stages 3 and 4 framework document*. Retrieved August 27, 2024, from https://www.gov.uk/government/publications/national-curriculum-in-england-secondary-curriculum

DfE (Department for Education). (2022). *Sustainability and climate change strategy*. Retrieved August 27, 2024, from https://www.gov.uk/government/publications/sustainability-and-climate-change-strategy/sustainability-and-climate-change-a-strategy-for-the-education-and-childrens-services-sy

Downing, D., & Watson, R. (2004). *School art: What's in it?: Exploring visual arts in secondary schools*. National Foundation for Educational Research.

Duverger, T. (2020). Degrowth: The history of an idea. *Encyclopédie d'Histoire Numérique de l'Europe*. Retrieved August 6, 2024, from https://ehne.fr/en/encyclopedia/themes/material-civilization/transnational-consumption-and-circulations/degrowth-history-idea

Eggleston, J. (2012). *The ecology of the school*. Taylor & Francis. Retrieved August 31, 2023, from https://www.perlego.com/book/1684654/the-ecology-of-the-school-pdf

Elias, J. (2016). *What kind of ecosystem is your school? Edutopia*. Retrieved August 12, 2024, from https://www.edutopia.org/blog/what-kind-ecosystem-your-school-maurice-elias

Fowkes, M., Fowkes, R., & Thomas, R. (2023). *Dear earth: Art and hope in a time of crisis*. Hayward Gallery Publishing.

Gadsby, H., & Bullivant, A. (2010). *Teaching contemporary themes in secondary education: Global learning and sustainable development*. Routledge.

Haraway, D. J. (2016). *Staying with the trouble*. Duke University Press.

Keegan, M. (2020). Hong Kong's terracotta tile army marches to the rescue for coral. *The Guardian*. Retrieved September 30, 2022, from https://www.theguardian.com/environment/2020/aug/26/hong-kongs-terracotta-tile-army-marches-to-the-rescue-for-coral

Long, R. (2023). Retrieved October 30, 2023, from http://www.richardlong.org

Mauch, C. (2019). Slow hope: Rethinking ecologies of crisis and fear. *RCC Perspectives, 1*. Retrieved February 12, 2024, from https://www.jstor.org/stable/26588127

Mollison, B. C., & Holmgren, D. (1978). *Permaculture 1: A perennial agriculture system for human settlements*. University of Tasmania.

OFSTED. (2022). *Guidance education inspection framework*. Retrieved September 19, 2023, from https://www.gov.uk/government/publications/education-inspection-framework/education-inspection-framework

Osterman, K. (2000). Students' need for belonging in the school community. *Review of Educational Research, 70*(3), 323–367.

Our Time on Earth. (2022). Retrieved February 15, 2024, from https://www.barbican.org.uk/whats-on/2022/event/our-time-on-earth

Pearsons. (2022). *The Pearson School Report 2022: Schools Today, Schools Tomorrow*. Viewed 30 September 2022 from https://www.pearson.com/engb/schools/insights-and-events/topics/school-report.html

Pittman, D., & Richmond, A. (2007). Academic and psychological functioning in late adolescence: The importance of school belonging. *The Journal of Experimental Education, 75*(4), 270–290.

Rugoff, R. (2023). *Dear earth: Art and hope in a time of crisis*. Hayward Gallery.

Scottish Government. (2023). *Learning for sustainability framework (LfS)*. Retrieved May 24, 2024, from https://www.gov.scot/publications/target-2030-movement-people-planet-prosperity/

Solnit, R. (2023). *Dear earth: Art and hope in a time of crisis*. Hayward Gallery.

Tanya, S., & Twyman, T. (2018). Considering visual arts practices at the secondary level: Extending cross-curricular conversations among secondary educators. *Art Education, 71*(2), 16–20.

TATE. (2023). *Olafur Eliasson's 'The Weather Project' filled the Turbine Hall at Tate in 2003*. Retrieved August 15, 2023, from https://www.tate.org.uk/whats-on/tate-modern/unilever-series/unilever-series-olafur-eliasson-weather-project/

Teach the Future. (2023). Retrieved August 14, 2023, from https://www.teachthefuture.uk

Trowell, J. (2010). Collaborative liberatory practices for global citizenship. In N. Addison, L. Burgess, J. Steers, & J. Trowell (Eds.), *Understanding art education: Engaging reflexively with practice*. Routledge.

UNICEF (The United Nations Children's Fund). (2015). *Children and the changing climate: Taking action to save lives*. Retrieved September 30, 2022, from https://www.unicef.org.uk/wp-content/uploads/2015/11/Unicef_2015childrenandclimatechange.pdf

Welsh Government. (2022). *Curriculum for Wales*. Retrieved May 24, 2024, from https://hwb.gov.wales/curriculum-for-wales

CHAPTER 10

What Is Collaborative Ecological Pedagogy?

Terri Newman

Introduction

Educating in art is a political act, which is seeking either to uphold present educational ideologies or to rupture them. Traditions in post-sixteen art and design education in England are deeply rooted in conventions of art-making and practices that constitute the known world. By this I mean the teaching and assessing of students' aptitude exists in relation to pre-established historical conventions for what art making is, and what it can be.

My practice-based research investigates how young artists learn differently when they engage in co-practice projects with their peers. My work is sited in the classroom and works with young artists aged 16–19 years old, studying art and design in England. I ask them to co-create an artwork with others over time. I co-research with the young artists that I teach, investigating how we learn and teach differently when we recognise the interconnected nature of artwork, meaning making, context and environment. My research is iterative and has evolved over ten years of

T. Newman (✉)
Goldsmiths, University of London, London, UK

M. Matthews (ed.), *Ecologies in Practice and Learning: Arts Interventions in the Earth Crisis*, Palgrave Studies in Educational Futures, https://doi.org/10.1007/978-3-031-82338-1_10

collaborative practice, working with young artists to develop methodologies for collaborative practices in the classroom.

In this chapter, I will articulate my practice-research that ruptures art and design pedagogies through the redistribution of learning relationships in the classroom, cultivating ecologies for learning. Beginning with my own experiences of the current curriculum model and its impacts upon the establishment of power relationships within the classroom and in assessment. I will propose collaborative art-making as an opportunity to explore the interconnected nature of our physical and social worlds.

My construction of an ecology of practice borrows from Stengers' positioning, as a 'tool for thinking through what is happening' (Stengers, 2013, p. 185), and in reference to collaborative pedagogies, 'what is happening' is that young artists are supported to recognise the entangled nature of their experiences. Collaborative pedagogies provide a tool through which young artists might examine the borders of their own knowledge building, constructing interconnected futures.

I will employ the viewpoints of Stengers (2013), Atkinson (2022) and Barad (2007) to articulate how ecologies in practice provide a tool through which young artists might examine the borders of their own knowledge building with others and their environment. I will then turn to Deleuze and Guattari (1987) to offer an illustration of how collaborative pedagogies are structured, enabling artists to build a sense of known self in relationship to the social and physical world. This chapter tracks how my practice-based research has allowed me to examine co-learning from a deeply embedded practitioner's standpoint as an artist/teacher/researcher (Thornton, 2013) and develop a tangible understanding of how co-practice in the curriculum offers something that lone exploration cannot create. In concluding this chapter, I will propose a methodology for collaborative pedagogies, beginning with the co-development of a criteria for success.

Collaboration, Power and the Curriculum

In our epoque, it is rare to find a teacher who will tell you that they actively discourage collaborative practices in all areas of their curricula; collaborative practices are encouraged in all subject areas and at all levels. In fact, in many of the schools I have worked in, collaborative learning opportunities in secondary schools in England will show you quiet classrooms, with students working independently, with support from the teacher. These practices are driven by the prioritisation of solo testing as a performance measure for schools. In the post-sixteen art and design classroom, the knowledge and opinions of the 'teacher' are often positioned as superior to the knowledge and opinions of the 'students', with a well-established bond between the attribution of power and the attribution of knowledge (Foucault, 1980). This model belongs to a history of instruction that presumes a fixed body of knowledge that can be known about a subject area that has ceased to develop, grow and evolve. It relies on the passive acceptance of a body of knowledge as 'correct'. In this model, the teacher is the one who teaches, and the students are the ones who are taught; the teacher is the one who knows, and the student does not (Freire, 1996). In this pedagogical model, the student's understanding of their subject is limited to what can be gathered from the guidance they receive from their teacher(s).

Ecologies and How Things Are Connected

My employment of ecologies builds upon Matthews' articulation of ecologies in practice as a collection of regenerative happenings, that support multiple viewpoints, nurturing cultures to support one another (Matthews, 2025). It is concerned with the entangled present as a point of divergence for multiple, yet to be understood, futures. Stengers' proposition is not a neutral one, returning to my positioning of the political in arts practice, for wherever we turn attention to the present, 'whatever understanding we have, comes to be included in this present anyway and cannot be separated from the understanding it generates' (Stengers, 2013, p. 185). This attention to how things are connected, rather than how they are, is also present in Deleuze and Guattari's (1987) multiplicities, which articulate the ways in which threads of understanding are thrown outwards, finding connections and divergencies—as they reach the threads of learning thrown out by others. The threads cannot be separate, they are entangled and so is the learning experience of the young artists that I work with.

Ecologies Are Inseparable from Context

I stated earlier that educating in art and design is a political act, but it is also one that is not static or fixed. It is a live act that cannot be separated from the time, place and company within which it grew. The artwork generated by the young artists that I work with articulates a very specific point within the lives of these 18-year-olds and the artists often choose complex issues to discuss, such as 'menstruation', 'LGBTQ+ rights', 'safe spaces' and 'feminism'. Making the artwork together gives them the opportunity to work through their own ideas with one another, as the artists of today building the contemporary practices of tomorrow. The artworks and art practices developed within the collaborative projects that I am building cannot be separated from the milieu, to borrow Stengers' (2013) use of the French term to refer to the environment, habitat or social setting.

Ecologies and Collaboration

My understanding of collaboration has two distinct strands of use for the word: one being combined actions in the division of labour and the other in the coming together of minds that build knowledge. De Wachter finds the merits of collaboration in that, 'It speeds things up because three minds and bodies operating together can operate at thrice the speed when required to do so' (De Wachter, 2017, p. 10). But seen as this, a collaboration might not serve a pedagogical function; rather, it is instrumental in achieving a set goal determined by pre-established boundaries. Miranda Matthews refers to ecologies in practice as producing 'more than a sum of its parts': considering this further, collaborative pedagogies provide space for each co-collaborator to reconsider their own ability to be employed within the aforementioned sum.

In order for a collaboration to be pedagogic, it must form ecosystems of interconnected practices that develop over time. Twyla Tharp, a 'professional collaborator', defines collaboration as 'people working together' (Tharp & Kornbluth, 2014, p. 4). She observes both good and bad collaborations. In 'good' collaboration, the collective can achieve more than the most talented individual may have been able to accomplish alone, Tharp's collaboration is an 'interchange-literally' where 'nothing forces change more dramatically than a new partnership' (Tharp & Kornbluth, 2014, p. 36), but it is more than this surely? Collaborative pedagogies provide a deeper experience than simply a fresh pair of eyes; do they not?

Johnson and Johnson ascertain that the cognitive activities and interpersonal dynamics 'only occur when students get involved in promoting each other's learning' (1999, p. 68). This is a process of rearticulation, feeding back into the group, rather an outside agency. For collaborative practices to nurture ecologies in practice, they must also address the systems of power within which they are formed. The linear journey described by the construction of the curriculum earlier reinforces the misunderstanding that arts practice at post-sixteen level is a journey with a beginning and an end, with a teacher who is the font of knowledge, whose gaze must be satisfied to achieve a high-level grade. Rather, in reality, learning, as we see in Barad, does not come from a singular teacher or information source; it exists as a series of affective events that transform established traditions of how young artists learn to develop their practice.

Entangled Futures and Multiplicities

Collaborative pedagogies do not attempt to build futures that repeat the practices of the past, recognising that educational policy in England is motivated by economic ambition that perpetuates social inequalities (Atkinson, 2022, p. 14). Collaborative practices approach the classroom at its divergencies that enable young artists to examine the borders of their experiences (Stengers, 2013, p. 184), with the borders of others, experimenting with the limits of their own understanding. An attention to how things are connected, rather than how they are, is also present in Deleuze and Guattari's (1987) multiplicities, which are defined by the 'lines of flight or deterritorialisation according to which they change in nature and connect with other multiplicities' (Deleuze & Guattari, 1987, p. 8), ever developing, changing and shifting which approach to practices as they 'diverge', generating understanding at these sites.

Ecologies in practice enable us to 'think through' pedagogies and to 'intra-act'. Barad's 'intra-action' moves away from interaction, in favour of recognising the ontological inseparability of interactions and our continual reconfigurations with the world (Barad, 2007). A pedagogy that begins with the intra-active nature of the lines of flight cast outwards is an approach that 'does not approach practices as they are' (Stengers, 2013, p. 184), recognising the continuously shifting renegotiations of the self in the social and natural world. It is an approach that in recognising the alive nature of these lines of flight illustrates a journey, not a destination. Collaborative pedagogies generate experiences that offer the opportunity to nurture ecologies in practice.

Asking young artists to pay attention to the lines of flight and the multiplicities that these generate, artists negotiate meaning with their collaborative partner as they develop their co-practice. The intra-actions that occur at these borders become a point of divergence at which new lines of flight might be traced. It is at these junctures that ecologies are nurtured. By considering the classroom a criss-crossed cartography of lines of flight (Deleuze & Guattari, 1987), we are not concerned with the reproduction of preconceived models for what art practice is or has been, but rather a 'world not yet known' (Atkinson, 2022, p. 27).

The Cartography of Ecologies

In the cartography of practice of many post-sixteen art and design classrooms, the young artists work silently; the cartography of teaching and learning zig zags between student and 'teacher' (Fig. 10.1).

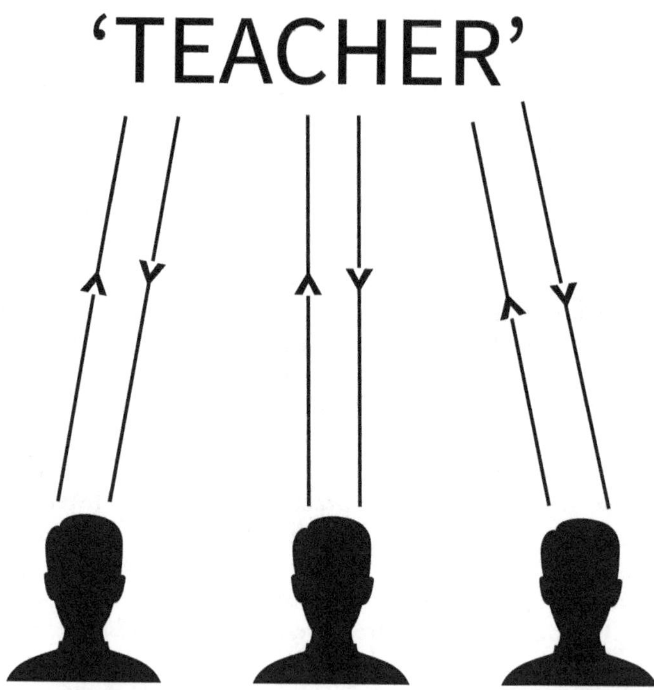

Fig. 10.1 Teaching and learning zig zags

In collaborative pedagogies, the cartography of practice is structured more like a rhizome, as Deleuze and Guattari employ the phrase (Deleuze & Guattari, 1987) re-articulating the role of the teacher to become co-researcher, with the young artists. 'A rhizome ceaselessly establishes connections between semiotic chains, organisations of power, and circumstances relative to the arts, sciences, and social struggles' (Deleuze & Guattari, 1987, p. 7). In a rhizome, there is no hierarchy of experience and knowledge; I am learning alongside the young artists, by positioning myself as co-researcher and undertaking my investigations *with* them. My knowledge about our shared area of investigation is more equal than when they are developing artworks that will ultimately be assessed by me. In this new model, there is no start or path to how knowledge is formed and understood. For a rhizomatic learning structure in collaborative pedagogy, this is how communities of learning are constructed, nurtured and maintained (Fig. 10.2). In collaborative pedagogies, the role of the teacher becomes the nurturing of such communities: learning to stand back and

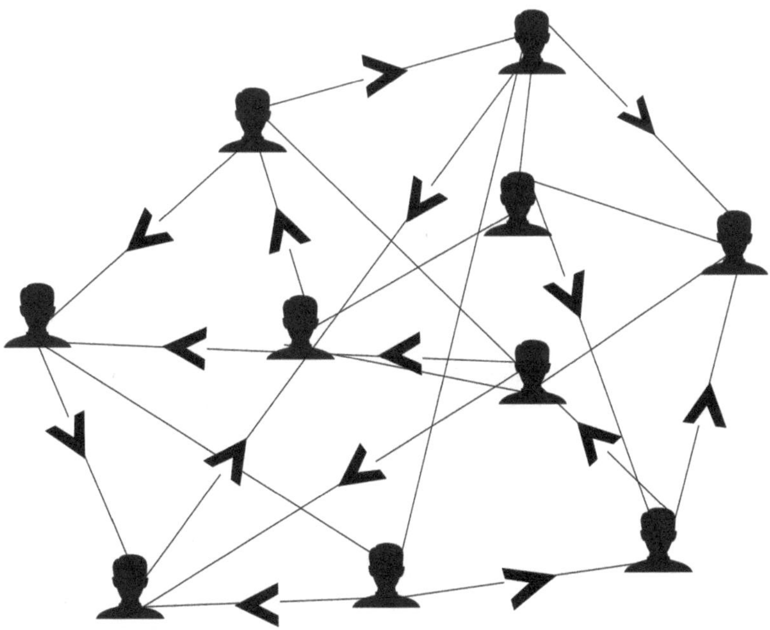

Fig. 10.2 Rhizomatic learning communities

trust the process when things are going according to plan. Communities of responsive learning begin to form under these conditions.

Collaborative pedagogies represent a shift in practice that restructures the learning relationship pathways. They move away from established practices in teaching and learning that rely on the subject content being prepared, hierarchised and presented by the teacher. Collaborative pedagogies support young artists to nurture ecological approaches to knowledge building, allowing them to 'practice' how to build understanding through intra-actions with other organisms and their environments.

A Methodology for Ecological Habits

My research employs practice methodologies which enable research insights to be drawn from engaging in the making of, or theorising of, creative work. Indicating that 'creative work in itself is a form of research' (Smith & Dean, 2009, p. 4). Practice-based methodologies in my research enable new understandings to be experienced in line with the development of ecologies. This is because they recognise and give space to explore the ways in which young artists examine the interconnections between their own perspectives, the perspectives of others and the visual language that they employ to express themselves and their perspectives (Sullivan, 2005, p. 43). Because of this, practices that enable reflection upon these experiences are crucial to the process. This is why the young artists that collaborate in these investigations do so armed with the reflective habits of the researcher (Schön, 1991). The young artists that participate in these projects are co-practice researchers engaged in finding out how they learn differently when they co-curate artwork with others.

Practice-based research also pays attention to the lack of neutrality experienced in qualitative research methodologies, recognising that we do not enter a research project as a neutral vessel. Rather, we take with us values, gender, politics ethnicity and so on; we also take our assumptions, categories, feelings and previous experiences (Coe et al., 2021, p. 73). In nurturing ecologies in practice, rather than experiencing this as a problematisation, practice-based research methodologies enable practitioners to examine how these intersectionalities impact experiences of learning. By asking each artist to keep a 'reflection log' after each encounter, I am able to track how the artists are impacted by their experiences and use these to guide 1–1 conversations with them and their artwork. The conversations enable us to gain new insights into collaborative practice by sharing our experiences of it. By building knowledge in this way, we are able to recognise the biases with which we approach the practice. Reynold and Mellor (2013) describe how, 'bodies are always "becoming" in relation to other bodies', always 'permeable to the influence (and imposition) of others, with porous borders and boundaries' (p. 33); collaborative pedagogies operate at this site, gaining insights into what it is to be a collaborative pedagogies practitioner.

The young artists are not taught *how* to collaborate but rather how to record their experiences so that they might employ them to make insights. The UK exam Board AQA asks students to reflect 'critically on work and progress' (AQA, 2017), referring to young artists recording how they might improve the impact of their creative practice. Young artists are asked to write these records using 'specialist language' and 'excellent English'. Collaborative pedagogies ask young artists to record their thoughts, feelings and responses to their collaborative practice as it develops, but place greater importance upon the impact of their actions with their collaborative partner, matter and materials, and the learning insights that these generate.

Criteria for Success in Collaborative Pedagogies

At the beginning of each project, the assessment criteria set forward by the exam board are abolished. The young artists must co-develop their assessment criteria for the project. By removing a top-down assessment model, the access to practice and assessment is structured rhizomatically (Deleuze & Guattari, 1987, pp. 1–29), giving the artists more agency and accountability for their projects. I pose questions such as, What does a good collaboration look like? How will you know if your collaboration has been a success? These negotiations often last hours and provide collective insight into the objectives of the project.

At the beginning stage of each project, the artists are given the space to understand through questioning, the true function of a practice to be built.

Young artist 1: It feels like we're not making rules for how to make art, it feels like we're making rules for being a good person.

Young artist 2: But if we're going to be making art with each other, then we need to be nice to each other.

Young artist 1: Yeah, but if I don't like your dress, then it's more nice to not tell you. I'm not gonna be like, oh I hate your dress that is being honest but it's not very kind.
Young Artist 3: No but if you're really collaborating, then you need to not worry about being nice, you gotta be honest.
Me: So, what's more important?
Young artist 2: To be respectful, you need to be honest, it all depends how you say it and what you mean.

(Transcribed Recording 2021)

The rule that the artists decided upon in the rhizomatic group was: 'Communicate with honesty and respect'. Collaborative pedagogies distribute agency and recognise conversations as above, illustrating how young artists have decided to prioritise the care with which we approach the social and natural world. When given the opportunity to develop their own assessment criteria, the young artists have chosen to prioritise the quality of the interaction that takes place, whilst working with matter, over the aesthetic quality of the object produced.

Developing their own success criteria is essential in that it gives the collaboration a framework to which they can refer, that is driven by themselves and not the exam board.

What I had not thought about was that these students had spent their lives institutionalised by criteria for success. By removing this entirely, I had taken the stabilisers off too early. It happened during the negotiations around the criteria, in which I saw the artists much more able to situate their own ideologies with that of working as part of a collective. The young artist visualises through the ideologies and ethics of others.

I employ practice-based research methodologies in my research because they enable young artists and me the opportunity to gain new understanding and knowledge, through collaborative arts practice and the artworks generated in the process (Candy & Edmonds, 2018). This is much like ecological approaches to practice, which pay particular attention to the interconnected nature of living organisms and their environment, recognising that objects, practice and meaning making cannot be disentangled.

Project Structure for Ecological Practice

The projects I have been working on have gone through four phases that were refined yearly from 2018 to 2023. In every iteration, I developed a collaborative practice that empowers young artists and allows them to feel confident in their work. When I started this work, I aimed to eliminate any elements of the traditional classroom environment that I felt reinforced the learner–teacher dynamic.

Each project follows Lewin's action research cycle, which involves identifying, analysing and addressing real-world problems. This cycle 'proceeds in a spiral of steps each of which is composed of a circle of planning, action, and fact-finding about the result of the action' (Lewin, 1946, p. 5). This approach allows young artists to see their collaborative arts practice as a research activity. It also gives them autonomy over their work and lets them experience the cyclical nature and ecosystems of creative practice. The phases of plan, present, develop and share are moved through cyclically as the project progresses.

Plan phase: During planning, young artists bring their ideas together through dialogue and mapping techniques to find harmony in their ideas. The intents and purposes of the collaboration are revisited and, if needed, renegotiated.

Present phase: In this phase, each collaborative pair shares their ongoing ideas with other members of the community involved in the project. This sharing session provides a critical gaze on the work. While I may facilitate the sessions initially, as confidence grows, the artists begin to lead their own meetings. At this stage, I must act as a mediator and mirror. It is where I must set aside any judgments I have and renegotiate my position as a teacher. Presentations are not given to me but to the other artists, and I encourage the artists to take on feedback provided by their peers.

Develop phase: This is where the artists bring their ideas into existence. My role here is not to interfere with the choices made by the artists but to ensure decisions and agreements are made together. It is important that questions are directed back to the collaborative pairing or group, as the artists should bring criticality to the practice.

Share phase: In this phase, the artists share their work with an audience, usually the rest of the community involved in the project.

Through these phases, the artists learn to work collaboratively, view their practice as a research activity, and gain confidence in their unique contributions to the creative process.

Conclusions

I intend with this chapter to introduce how collaborative projects in the classroom nurture ecologies in practice. Employing ecologies as a 'tool for thinking through' how collaborative learning partnerships enables young artists to explore the interconnected nature of their artistic practice, their social and material worlds.

I began this chapter by articulating the entangled nature of power and knowledge within the post-sixteen classroom and the ways in which this reinforces neoliberal ideologies of individualised success, lone working and isolated assessment. I have outlined how these practices assume that what can be known or understood about a subject is fixed. In this chapter, I have presented ecologies as a series of interconnected, regenerative happenings that cannot be separated from the meaning they generate.

In this discussion of my research into collaborative pedagogies, I have stated how I believe co-creating artworks in collaboration with peers enables young artists to build practices for interconnected futures with one another and their environments. These practices enable us to develop social bonding, community building and a recognition of our responsibility to the physical world.

I have aimed to articulate the experience of examining the divergencies at the borders of our judgments and understanding. I have also explained how my practice nurtures these divergencies in the post-sixteen classroom. Ecologies in practice draw attention to how things are interconnected and how these affections can build innovative futures. My research proposes collaborative pedagogies as sites at which ecologies can be nurtured and asks how we can learn more about the world, our position within it, and our ability to cause affect.

Over ten years, I have co-developed a methodology for collaborative pedagogies that has been co-shaped with young artists. It is a way of working that enables critical reflection of the learning structures that define what can be understood through the study of art and design. Working in this way allows us to pay particular attention to how hierarchies of knowledge are formed and the interconnections between our physical and social worlds. These ways of working encourage us to consider relegating the idea of a tortured artist alone in their studio and consider the notion that creative arts practice is not founded upon anthropocentric ideologies. This chapter proposes that collaborative pedagogies enable shifts in power relationships in learning, turning attention to multiplicities and how they are pedagogical.

Competing Interests (4) The author has no conflicts of interest to declare that are relevant to the content of this chapter.

References

Atkinson, D. (2022). *Pedagogies of taking care*. Bloomsbury Publishing.

AQA. (2017). *Art and design*. Retrieved August 7, 2024, from https://www.aqa.org.uk/subjects/art-and-design/as-and-a-level/art-and-design-7201

Barad, K. (2007). *Meeting the universe halfway: Quantum physics and the entanglement of matter and meaning*. Duke University Press.

Candy, L., & Edmonds, E. A. (2018). Practice-based research in the creative arts: Foundations and futures from the front line. *Leonardo, 51*(1), 63–69.

Coe, R., Waring, M., Hedges, L. V., & Arthur, J. (2021). *Research methods & methodologies in education* (3rd ed.).

Deleuze, G. & Guattari, F. (1987). *A thousand plateaus: Capitalism and schizophrenia*. Translated by B. Massumi. Bloomsbury.

De Watcher, E. (2017). *Co-art: Artists on creative collaboration*. Phaidon.

Freire, P. (1996). *Pedagogy of the oppressed*. Penguin.

Foucault, M. (1980). *Power/knowledge: Selected interviews and other writings, 1972–1977*. Pantheon Books.

Johnson, D. W., & Johnson, R. T. (1999). Making cooperative learning work. *Theory Into Practice, 38*(2), 67–73.

Lewin, K. (1946). Action research and minority problems. *Journal of Social Issues, 2*(4), 34–46.

Matthews, M. (2025). Introduction. In M. Matthews (Ed.), *Ecologies in practice and learning: Arts interventions in the Earth Crisis*. Palgrave Macmillan.

Reynold, E., & Mellor, D. (2013). Deleuze and Guattari in the nursery: Towards an ethnographic multi-sensory mapping of gendered bodies and becoming. In R. Coleman & J. Ringrose (Eds.), *Deleuze and research methodologies* (pp. 23–42). Edinburgh University Press.

Schön, D. (1991). *The reflective practitioner: How professionals think and act*. Avebury.

Smith, H. & Dean, R. T. (2009). *Practice-led research, research-led practice in the creative arts*. Edinburgh: Edinburgh University Press.

Stengers, I. (2013). Introductory notes on an ecology of practices. *Cultural Studies Review, 11*(1), 183–196.
Sullivan, G. (2005). *Art practice as research.* Sage.
Tharp, T., & Kornbluth, J. (2014). *The collaborative habit: Life lessons for working together.* Simon & Schuster.
Thornton, A. (2013). *Artist, researcher, teacher: A study of professional identity in art and education.* Intellect.

CHAPTER 11

Ecological Pedagogies for Planetary Health: New Ways of Thinking in Higher Education

Kevin Campbell Davidson

Introduction

Young people are growing up in a time when the warming of global climates threatens life on Earth and the sustainability of the Earth's biosphere (IPCC, 2023). In response to these forecasts, 2355 local authorities in 40 counties have declared a state of climate emergency (CED, 2024). 'Climate emergency' and 'climate crisis' are spiking in academic literature, from 32 papers in the Web of Science database in 2015 to 862 papers in 2022 (Osaka, 2023). The anthropogenic causes of climate change, such as increases in greenhouse gas emissions and deforestation, are resulting in extreme weather events, an increase in natural disasters, loss of biodiversity, rising sea levels and decrease in crop productivity (IPCC, 2023; Shivanna, 2022). Given these foreboding signs about the future, it is perhaps not surprising that young people throughout the world are experiencing eco-anxiety and future-related uncertainty (Hickman et al., 2021).

K. Campbell Davidson (✉)
University of London, London, UK
e-mail: k.davidson@gold.ac.uk

M. Matthews (ed.), *Ecologies in Practice and Learning: Arts Interventions in the Earth Crisis*, Palgrave Studies in Educational Futures, https://doi.org/10.1007/978-3-031-82338-1_11

Educators are increasingly concerned about how to respond to young people's experiences of eco-anxiety and the role of ecological pedagogies in addressing these complex emotions (Verlie et al., 2021; Pihkala, 2020). Research with educators in the UK and Canada suggests that young people experiencing eco-anxiety may be supported by activities which improve nature connection (Edwards et al., 2023). This may include 'cognitive' nature connection activities which promote ecological literacy and an understanding of the embeddedness of humans within natural systems (Beery et al., 2023). Since human–nature connectedness has shown to be positively correlated with pro-environmental behaviour (Muhr, 2020), ecological literacy may be an important driver of behavioural change as well as an antidote to eco-anxiety.

Human behavioural change is essential for mitigating the effects of anthropogenic causes of climate change (Bergquist et al., 2023). However, these environmental problems and the possible solutions are not only due to what humans do but also how we think (Deleuze & Guattari, 1988). As centres of learning and development, higher education institutions have the potential to change human thinking and behaviour. In order to do this, they must go beyond a 'greening' of education which simply adds climate education content to existing university curricula, towards more fundamental shifts in 'onto-epistemology'—which is a term for ways of being and thinking (Carstens, 2016). This paradigm shift will challenge 'fictions of separation' between humans and non-human beings, recognising our relationality and embeddedness in a shared world (Carstens, 2016, p. 265). Students may be confronted with intellectual uncertainty as they engage with material which is uncomfortably strange, in the sense that simplistic binary assumptions (e.g. human/nature) are challenged and the familiar becomes unfamiliar (Barnett, 2000). This kind of 'epistemological stretching' (Barrett et al., 2017)—a valuable expansion of our relation to knowing—needs to be supported by pedagogical theory.

Posthumanist theories share a fundamental concern for rethinking human relationships through a de-centring of the human subject, recognising the agency and value of non-human entities and the environment (Braidotti, 2013; Gruen, 2015; Haraway, 2016). Duobliene and Vaitekaitis (2021) demonstrate how posthumanist-inspired critical thinking can lead to a re-articulation of animal–nature–human relationships embedded within current educational curricula, challenging 'human exceptionalism' which characterises humans as superior to other species, which are portrayed as mainly a resource for humans with no intrinsic value. A transition from human-centred pedagogies to posthumanist ecological pedagogies will promote environmental ethics and responsible stewardship of the Earth (Jickling & Sterling, 2017). By integrating ethical considerations into education, ecological pedagogies can help cultivate a generation of environmentally conscious individuals who are aware of their embeddedness within natural systems and equipped to address the complex challenges posed by climate change (Gruen, 2015). This shift towards an 'ecological consciousness', which recognises the interdependence of human and planetary sustainability, has been described as the education which is needed for future survival (Common Worlds Research Collective, 2020). The severity of the climate emergency demands that we change not just what we teach but the way we teach.

Systems Thinking and Multiple Ways of Knowing

In moving from human-centred to ecological pedagogies, non-human beings are recognised as active communicating agents within living systems, rather than merely isolated objects to be studied. Their agency is interconnected with human lives through food production, energy consumption, and use of natural resources among much else. This is why it has been suggested that an appreciation of complex interconnections within living systems, or 'systems thinking', is an important foundation for learning about sustainability (Sandri, 2013, p. 811). When students become familiar with the characteristics and behaviour of complex systems, they are more able to understand the ways in which human activity impacts natural ecosystems. The cognitive habits of a systems thinker include a propensity to see the whole picture and see oneself as part of the system, as well as an ability to change perspectives, look for interdependencies, find unanticipated consequences, and hold the tension of paradox without seeking quick resolution (Meadows et al., 2016). To draw upon posthumanist terminology, students can enter liminal spaces where common assumptions are challenged via 'transversal thinking', moving between different frames of reference to understand reality as a multi-layered whole made from interlocking parts (Palmer & Panayotov, 2016). This ecological understanding can be supported by teaching through relevant threshold concepts.

Threshold Concepts

Given the challenges of adopting different ways of thinking and doing, Duobliene and Vaitekaitis (2021) recommend a soft transition in educational approaches from human-centred to ecological pedagogies. It may be helpful to consider 'threshold concepts' (Barradell, 2013) which invite students towards the critical reflection and contemplation required for new ways of thinking. Threshold concepts are distinguished from core concepts on the basis that they open up a qualitatively different understanding of the subject matter (Meyer & Land, 2006). Threshold concepts can be both transformative, taking students into a liminal space which prompts onto-epistemological change (Quinlan et al., 2013), and troublesome, as deeply ingrained assumptions are challenged (Sterling, 2010). From this perspective, the role of higher education institutions is to introduce threshold concepts which guide students through this transition towards new ways of understanding the human's place in wider ecologies.

Multiple Ways of Knowing

Barrett et al. (2017) explored the transition towards ecological awareness through a piece of action research with students and educators, deriving a set of six threshold concepts which are outlined as follows: (1) there are different ways of knowing; (2) we can communicate with non-human nature and non-human nature can communicate with us; (3) knowing is relational; (4) Transrational intuition and embodied knowing is a valuable and valid way of knowing; (5) worldview is the lens through which we experience reality; (6) power of dominant beliefs supports and/or undermines particular ways of knowing and being as in/valid (Barrett et al., 2017, p. 136). In summary, these threshold concepts invite students to challenge dominant ideas about human-centric ways of knowing which create a separation between humans and non-human beings.

Arts-Based Practice

Scholars of arts-based practice recognise art-making as a way of knowing (McNiff, 2011). In articulating the idea that a complex understanding of the world can be reached through non-discursive methods (Langer, 1953), such as pictures, music or dance, the implicitly anthropocentric dominance of verbal discourse which exists in Western academic traditions may be challenged (Barone & Eisner, 2012; Roulstone et al., 2011). The binary human/nature separation may be challenged by recognising the fundamental role of sensory experience in arts-practice, which has been described as 'knowing-in-action: a knowing-with and knowing-through' (Feld, 2015, p. 12). Faye (2023) builds an argument that the sensory approaches to developing knowledge employed by non-human animals are not radically different from those of human sensory knowledge. For example, just as humans may visually imagine how to get from A to B without necessarily being able to describe the route in linguistic terms, equally, this kind of image-based knowledge is likely to be possessed by non-human animals. Faye recognises that the embodied knowledge shared by all animals, including humans, is acquired by doing, by practice rather than by reflection, and although it may be partially expressible as language, it may also exist beyond language. The following section goes on to consider how a games-based arts practice pedagogy can be employed in higher education settings to develop ecological awareness, multiple ways of knowing and 'systems thinking' capacities.

Systems Thinking Games as Arts Practice Pedagogy

Arts practice research enables researchers to share the ways of knowing that emerge in arts practice, including intuitive, embodied, tacit, affective and sensory ways of knowing (Bulley & Şahin, 2021). As well as being used by researchers to generate research findings, arts practice research may be used as a pedagogical approach where arts-based methods are integrated with theoretical learning (Hunter & Frawley, 2023; Bayley, 2018). The learning and insights gained through the process of research may be of equal or even greater value than the research findings. For example, participants engaged in social land art projects made their own art installations whilst reflecting upon their relations with the natural setting, revealing deep emotional connections to specific landscapes through group discussions and participant observations (Riechers, 2019). Arts-based research such as this may advance students' appreciation of different ways of knowing, including intuitive, embodied and preverbal ways of knowing which can be shared with non-human nature, which are threshold concepts in ecological thinking (Barrett et al., 2017, pp. 137–139).

This approach to developing knowledge and understanding through multiple embodied, sensory and performative methods has been described as 'sensory literacy' (Matthews, 2021) and has the potential to tap into an emotional form of human-nature connectedness which lies 'beyond words' and which has the potential to promote sustainable behaviour (Muhr, 2020). For example, a participatory arts project which combined movement exercises, guided drawing exercises and reflective discussion revealed nuances of emotional and human-nature connectedness which would likely not have been revealed by verbal engagement alone (Muhr, 2020, p. 252). In this way, arts practice invites students to look from multiple perspectives and is potentially supportive of systems thinking capacities, such as the ability to change perspectives (Bulley & Şahin, 2021).

Games have been recognised as a form of pedagogical arts practice which generates collaborative arts-making, interpersonal discourse and reflection (Cox, 2014). Systems Thinking Games (STGs) are experiential learning activities which strategically guide students towards systems thinking capacities and raise awareness of the interconnections between human activity and the lives of others (Meadows et al., 2016). Through the conscious use of the body, STGs acknowledge the value of embodied ways of knowing and the subconscious intelligence which lives in the human body (Claxton, 2012). For example, in the 'Arms Crossed' game students are invited to cross their arms in the opposite way to their habitual manner (e.g. right arm on top when usually they put left arm on top) which stimulates intuitive inquiry through active use of the body (Meadows et al., 2016). This gives students an opportunity to experience an embodied appreciation of the process of disrupting habitual behaviour in favour of new forms of behaviour. In the reflective discussion which follows, students will be invited to consider their habitual behaviours, including behaviours they may wish to change. This discussion may raise specific examples about human behavioural habits which impact upon environmental systems and lead onto wider discussions around the broader human systems which promote and constrain individual choices and associated ethical implications.

Other STGs are designed to highlight to students their unconscious assumptions about the nature of reality, inviting them to step outside of their own viewpoint on a complex problem and look at other worldviews (Barrett et al., 2017, pp. 139–140). To use the language of a systems thinker, students may come to see themselves as part of a system and develop an ability to change perspectives. In considering the many interlocking parts of a multi-layered system, students may be drawn towards 'transversal thinking' (Deleuze & Guattari, 1988). This process has the potential to generate learning environments where students are discussing issues which are relevant and meaningful in their lives, whilst also developing systems thinking capacities and learning about ecological sustainability. STGs should be followed by open-ended peer-to-peer discussions, as opposed to teacher-led sessions centred around learning objectives, and this can be supported by reflective pedagogical practices such as reciprocal teaching.

Reciprocal Teaching

Reciprocal Teaching (RT) is a process whereby students mutually clarify their levels of understanding and in doing so recognise that they have knowledge to bring to the group (Palinscar & Brown, 1984). Originally developed as a reading intervention, it has been applied in other situations, including higher education (Gilbert, 2017). In practical terms, students form into small groups or pairs and discuss a series of prompts in relation to a specific stimulus material, such as an STG. The prompts invite students to consider what they have understood from the material, what they have not understood, and what other questions have arisen for them. Each small group/pair then feeds back one or two key points from their discussion to the whole group, who may ask questions or develop the discussion further. In this way, students have the opportunity to develop knowledge through social, affective and relational occurrences. These "knowledge-ing" interactions may encourage students to take risks by overcoming a potentially limiting concern to find the one 'right' answer (Taylor & Harris-Evans, 2016, p. 13). For example, in my practice, I have observed that students may be more willing to express intuitive, emergent and 'un-thought-through' ideas in pairs than when speaking in front of a larger group.

These ideas may be developed and clarified by peers who support each other in finding the words and confidence to express themselves. This process implicitly highlights that 'knowing is relational' (Barrett et al., 2017, p. 138), since each RT pair will generate different responses to the same source material due to the dialogical nature of knowledge-ing. Deleuze and Guattari (1988) describe this as rhizomatic forms of knowing, which can spread in any direction, move through different levels and through multiple channels in a non-hierarchical way. In this way, RT decentralises decision-making about which knowledge is valued, reducing the 'power of dominant beliefs' (Barrett et al., 2017, p. 140), such as those of a lecturer. Since the knowledge of RT emerges from student interactions, the lecturers' role changes from 'knowledge-deliverer' to 'rhizomatic-facilitator' and the social influence of the members of the group is given greater space to thrive.

Behavioural Change

There is evidence to suggest that practices such as RT may be effective to promote the human behavioural change necessary for the mitigation of the effects of anthropogenic causes of climate collapse. In a meta-review of research evidence on field interventions aiming to promote climate change mitigation behaviours, it was found that interventions using social influence have the strongest effects on improving sustainable behaviour, since individuals conform to social norms to gain the approval of others (Bergquist et al., 2023). On this basis, giving students the opportunity to hear the opinions of their peers about issues which are relevant and meaningful to their lives may be a more powerful driver of behaviour change than traditional approaches where a lecturer builds a case for behaviour change based on facts. Students who hear a range of responses to pedagogical arts practice, such as STGs, may find themselves re-evaluating their understanding and beliefs.

Systems Thinking Games in Practice

Having established the theoretical basis for Systems Thinking Games as a form of pedagogical arts practice which has the potential to develop sensory literacy, systems thinking capacities and multiple ways of knowing, suggestions will now be given as to how these activities may be facilitated in an integrated process of learning. This will be demonstrated through an STG, *The Empty Chair*, which I have developed with students on the BA Educational Studies programme at Goldsmiths University of London.

The Empty Chair

The game is played as follows. A group of students sit in chairs which are spread throughout the learning space. Additionally, there is one empty chair on the other side of the room from the facilitator. The facilitator's objective is to sit down in the empty chair, whilst the students' mutually shared objective is to prevent this from happening. In order to clarify their role, the facilitator may wish to introduce themselves as a 'dumb' or

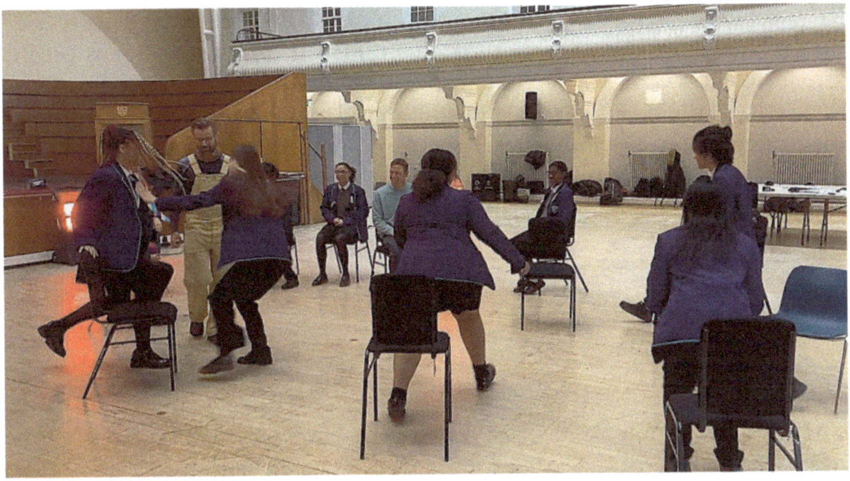

Fig. 11.1 Empty chair game

'pre-programmed' robot who follows an algorithm to head to the nearest empty chair. When the game commences, the facilitator walks slowly in the direction of the empty chair with the intention to sit down in it. The group may prevent this from happening by one or more of the students standing up and sitting down in the empty chair. This of course leaves another empty chair to which the facilitator now slowly moves towards. The game continues until the facilitator sits down or until the end of the game is announced (Fig. 11.1).

Reflective Practice

Over successive attempts, most student groups become increasingly successful in preventing the facilitator from sitting down by developing collaborative strategies through reflective practices such as RT. A number of different insights may emerge. For example, one pair may notice that a person who is far away from the empty chair is strategically best placed to move over to protect it, since this leaves the new empty chair very far away from the facilitator, allowing time for the next protection to be made. This realisation relates directly to a systems thinking capacity to see the whole picture rather than just the part nearby the action. Another RT pair may notice that when one player simply follows the facilitator around the room, they will usually be able to protect the empty chair effectively. This realisation relates directly to another systems thinking capacity, to focus on the interconnections between elements (the facilitator) rather than the effects themselves (the chairs). Yet another RT pair may notice that having more than one person moving around the room makes it more difficult for the facilitator to sit down, even though there are a greater number of empty chairs available to them. This discovery, which may initially appear counterintuitive, emerges from the fact that the greater number of potential protectors outweighs the increased availability of empty chairs and relates directly to the systems thinking capacity to hold the tension of paradox.

Applying Systems Thinking for Behavioural Change

The insights gained about how to play the game effectively may then be related to subject-specific learning and connected with wider ecological thinking. For example, in educational studies, students studying creative learning have connected embodied experiences arising from *The Empty Chair* with theoretical knowledge about the 'distributed' nature of creative collaboration between audiences, materials and embodied action (Barrett et al., 2021). This has led to considerations about the role of living systems in creative media, highlighting an opportunity for students to change behaviour by streaming music in a way which credits the contribution of the more-than-human beings to the work (Sounds Right, 2024).

Muhr (2020) suggests that the emotional and philosophical connections to nature which are nurtured by arts-based methods may constitute 'deep leverage points' which influence the underlying goals and values of a system such as higher education. For example, educational studies which use arts-based methods to develop emotional and philosophical connections to nature may go on to promote the well-established benefits of outdoor learning for combatting eco-anxiety through 'experiential' nature connection (Beery et al., 2023). They may take a critical position on the systems-level factors which present barriers to outdoor learning, such as teacher training, school design and assessment practices (Beyond Barriers, 2010). In this way, the promotion of systems thinking capacities may affect students' values and pro-environmental behaviour and may empower arts educators to take an ecological way of thinking into their future work.

As teaching shifts from an anthropocentric to an ecological paradigm, students may place greater value in sensory literacy, integrate their local environment in their learning and use the process of education to act as guardians of the legal rights of non-human actors such as rivers (Pecharroman, 2018). An appreciation of complex systems and underlying interspecies interdependency can promote ecological consciousness in every subject discipline and lead to a consideration of the impacts of the modern ways of living upon global environmental collapse. By equipping students with a different way of thinking about the human's place within the world, higher education can act as a driver of human behavioural change.

Conclusion

Higher education institutions are well placed to instigate the changes in human behaviour which are necessary to mitigate anthropocentric causes of global warming and associated environmental collapse. Pedagogical practices in higher education institutions need to change to support young people experiencing eco-anxiety and equip them to face an increasingly complex and uncertain world. Cognitive human-nature connectedness can be fostered through systems thinking and an appreciation of multiple ways of knowing. The paradigm shift from human-centred to ecological consciousness can challenge dualist assumptions which separate humans from the natural world and recognise the embeddedness of humans within wider ecological systems. In turn, this may widen the view of the kind of knowledge which is considered valid in higher education programmes.

In order to make this transformative and sometimes troublesome transition, threshold concepts may serve as guideposts in the implementation of novel approaches. Pedagogical arts-based practices such as Systems Thinking Games, supported by reflective processes such as Reciprocal Teaching, may be used in combination to help students to explore new ideas. This chapter invites further research and discussion about the role of higher education institutions and practices in responding to the climate crisis and supporting the future lives of young people.

Acknowledgement The author would like to gratefully acknowledge the contributions of Tom Mansfield of Pale Blue in the development of the ideas within this chapter.

Competing Interests (4) The author has no conflicts of interest to declare that are relevant to the content of this chapter.

References

Barradell, S. (2013). The identification of threshold concepts: A review of theoretical complexities and methodological challenges. *Higher Education, 65*, 265–276.

Barnett, R. (2000). University knowledge in an age of supercomplexity. *Higher Education, 40*, 409–422.

Barone, T., & Eisner, E. (2012). *Arts based research*. Sage.

Barrett, M. J., Harmin, M., Maracle, B., Patterson, M., Thomson, C., Flowers, M., & Bors, K. (2017). Shifting relations with the more-than-human: Six threshold concepts for transformative sustainability learning. *Environmental Education Research, 23*(1), 131–143.

Barrett, M. S., Creech, A., & Zhukov, K. (2021). Creative collaboration and collaborative creativity: A systematic literature review. *Frontiers in Psychology, 12*. https://doi.org/10.3389/fpsyg.2021.713445

Bayley, A. (2018). *Posthuman pedagogies in practice: Arts based approaches for developing participatory futures*. Springer Verlag.

Beery, T. A., Stahl Olafsson, S., Gentin, M., Maurer, S., Stålhammar, C., Albert, C., & Bieling, A. (2023). Disconnection from nature: Expanding our understanding of human–nature relations. *People and Nature, 5*(2), 470–488.

Bergquist, M., Thiel, M., Goldberg, M. H., & van der Linden, S. (2023). Field interventions for climate change mitigation behaviors: A second-order meta-analysis. *Proceedings of the National Academy of Sciences of the United States of America, 120*(13), e2214851120. https://doi.org/10.1073/pnas.2214851120

Beyond Barriers. (2010). *Beyond barriers to learning outside the classroom in natural environments*. Retrieved June 18, 2024, from http://publications.naturalengland.org.uk/file/6621187224371200

Braidotti, R. (2013). *The posthuman*. Polity Press.

Bulley, J., & Şahin, Ö. (2021). *Practice research - report 1: What is practice research? And report 2: How can practice research be shared?* Practice Research Advisory Group UK (PRAG-UK).

Carstens, D. (2016). The anthropocene crisis and higher education: A fundamental shift. *South African Journal of Higher Education, 30*(3), 255–273. Retrieved August 7, 2024, from https://doi.org/10.20853/30-3-650

CED (Climate Emergency Declaration). (2024). *Climate emergency declaration*. Retrieved August 7, 2024, from https://climateemergencydeclaration.org/climate-emergency-declarations-cover-15-million-citizens

Claxton, G. (2012). Turning thinking on its head: How bodies make up their minds. *Thinking Skills and Creativity, 7*(2), 78–94.

Common Worlds Research Collective. (2020). *Learning to become with the world: Education for future survival*. UNESCO. Retrieved August 7, 2024, from https://unesdoc.unesco.org/ark:/48223/pf0000374032

Cox, J. (2014). Role-playing games in arts, research and education. *International Journal of Education Through Art, 10*(3), 381–395.

Deleuze, G., & Guattari, F. (1988). *A thousand plateaus: Capitalism and schizophrenia*. Athlone Press.

Duobliene, L., & Vaitekaitis J. (2021). *Journal of Futures Studies, 26*(2), 37–50.

Edwards, R. C., Brendon, M., Larson, H., & Clayton, S. (2023). Navigating eco-anxiety and eco-detachment: Educators' strategies for raising environmental awareness given students' disconnection from nature. *Environmental Education Research, 30*(6), 864–880.

Faye, J. (2023). *The biological and social dimensions of human knowledge*. Palgrave Macmillan.

Feld, S. (2015). Acoustemology. In D. Novak & M. Sakakeeny (Eds.), *Keywords in sound* (pp. 12–21). Duke University Press.

Gilbert, F. (2017). *Reciprocity and reading*. Retrieved June 17, 2024, from https://research.gold.ac.uk/id/eprint/22466/

Gruen, L. (2015). *Entangled empathy: An alternative ethic for our relationships with animals*. Lantern Books.

Haraway, D. J. (2016). *Staying with the trouble: Experimental futures*. Duke University Press.

Hickman, C., Marks, E., Pihkala, P., Clayton, S., Lewandowski, R. E., Mayall, E. E., Wray, B., Mellor, C., & van Susteren, L. (2021). Climate anxiety in children and young people and their beliefs about government responses to climate change: A global survey. *The Lancet. Planetary Health, 5*(12), e863–e873. https://doi.org/10.1016/S2542-5196(21)00278-3

Hunter, L., & Frawley, E. (2023). Engaging students using an arts-based pedagogy: Teaching and learning sociological theory through film, art, and music. *Teaching Sociology, 51*(1), 13–25.

IPCC (Intergovernmental Panel on Climate Change). (2023). Climate change 2023: Synthesis report. In H. Lee & J. Romero (Eds.), *Contribution of working groups I, II and III to the sixth assessment report of the IPCC* (pp. 35–115) Geneva. Retrieved August 7, 2024, from https://doi.org/10.59327/IPCC/AR6-9789291691647.

Jickling, B., & Sterling, S. (2017). *Post-sustainability and environmental education: Remaking education for the future*. Palgrave Macmillan.

Langer, S. K. (1953). *Feeling and form*. Charles Scribner's.

Matthews, M. (2021). Sensory literacy and empowerment in art and design. *International Journal of Art & Design Education, 40*(3), 581–598.

Meadows, D., Booth Sweeney, L., & Martin Mehers, G. (2016). *The climate change playbook*. Chelsea Green Publishing.

McNiff, S. (2011). Artistic expressions as primary modes of inquiry. *British Journal of Guidance & Counselling, 39*(5), 385–396.

Meyer, J. H. F., & Land, R. (2006). Threshold concepts and troublesome knowledge: An introduction. In J. H. F. Meyer & R. Land (Eds.), *Overcoming barriers to student understanding* (pp. 3–32). Routledge.

Muhr, M. (2020). Beyond words—The potential of arts-based research on human-nature connectedness. *Ecosystems and People, 16*(1), 249–257.

Osaka, S. (2023, October 30). Why many scientists are now saying climate change is an all-out 'emergency'. *Washington Post.* Retrieved August 7, 2024, from https://www.washingtonpost.com/climate-environment/2023/10/30/climate-emergency-scientists-declaration/

Palinscar, A., & Brown, A. (1984). Reciprocal teaching of comprehension-fostering and comprehension-monitoring activities. *Cognition and Instruction, 1*(2), 117–175.

Palmer, H., & Panayotov, S. (2016, September 13). Transversality. *New Materialism.* Retrieved August 7, 2024, from https://newmaterialism.eu/almanac/t/transversality.html

Pihkala, P. (2020). Anxiety and the ecological crisis: An analysis of eco-anxiety and climate anxiety. *Sustainability, 12*(19), 7836.

Pecharroman, C. L. (2018). Rights of nature: Rivers that can stand in court. *Resources, 7*(1), 13.

Riechers, M. (2019). Stories of favourite places in public spaces: Emotional responses to landscape change. *Sustainability, 11*(14), 3851.

Roulstone, S., Law, J., Rush, R., Clegg, J., & Peters, T. (2011). *Investigating the role of language in children's early educational outcomes.* Department for Education. Retrieved August 7, 2024, from https://www.gov.uk/government/publications/investigatingthe-role-of-language-in-childrens-early-educational-outcomes

Quinlan, K. M., Male, S., Baillie, C., Stamboulis, A., Fill, J., & Jaffer, Z. (2013). Methodological challenges in researching threshold concepts: A comparative analysis of three projects. *Higher Education, 66*(5), 585–601.

Sandri, O. J. (2013). Threshold concepts, systems and learning for sustainability. *Environmental Education Research, 19*(6), 810–822.

Shivanna, K. R. (2022). Climate change and its impact on biodiversity and human welfare. *Proceedings of the Indian National Science Academy. Part A, physical sciences, 88*(2), 160–171.

Sounds Right. (2024). Retrieved August 7, 2024, from https://www.soundsright.earth

Sterling, S. (2010). Transformative learning and sustainability: Sketching the conceptual ground. *Learning and Teaching in Higher Education, 5*(11), 17–33.

Taylor, C., & Harris-Evans, J. (2016). Reconceptualising transition to higher education with Deleuze and Guattari. *Studies in Higher Education, 43*(7), 1254–1267.

Verlie, B., Clark, E., Jarrett, T., & Supriyono, E. (2021). Educators' experiences and strategies for responding to ecological distress. *Australian Journal of Environmental Education, 37*(2), 132–146.

PART IV

Sustainable Arts Practice: Arts Educators in Museums, Galleries and Community Spaces

Inclusive participation in sustainable curating, discursive processes for artist educators working with galleries, and the expansion of ecological learning in museums, galleries and community spaces form connecting threads in this section. Authors raise problem spaces such as the need for cultural organisations to shift their mission statements and modes of activity, to enable participants to question and vocalise issues. Methods put into action by the contributing authors astutely address causal factors in the current conditions of climate change such as the loss of biodiversity and effects of extractivist capitalism.

Bridget McKenzie discusses how a 'possitopian' approach is taken by Climate Museum UK, a distributed museum that has collected artefacts in locations around the UK as well as online resources. CMUK shares up-to-date information and assists families, school students and community groups to understand Earth Crisis issues. McKenzie asks probing questions about the kinds of learning experiences that are suitable for addressing the complexities of the crisis through arts practice. She presents a trauma-sensitive approach that is attuned to the needs of communities and individuals, drawing upon projects that CMUK has delivered, to help participants visualise possibilities for reparations to turn around the causes and impacts of the crisis.

Trish Scott analyses how contemporary galleries can enable a sustainable and inclusive relation to their local communities, reducing the operational carbon footprint via the ecological perspective of 're-wilding' the gallery. Relating to the 'Residents' programme at Goldsmiths Centre for

Contemporary Art in London, UK, Scott re-positions the gallery as a pluriverse of lateral power relations. Audiences from marginalised groups and their local ecosystems, as well as arts educators, are viewed more equally in relation to artists, curators and those who are already at ease with gallery cultures. This chapter offers supportive concepts for more galleries to include a re-wilding space.

Penny Hay and Andrew Amondson focus their contribution to this collection on the living installations they have created in an annual arts festival called Forest of the Imagination in the city of Bath, UK. Calling upon play and creative engagement as forms of self-expression that assist widening participation in learning about arts ecologies, Hay and Amondson present art spaces that act as both a sanctuary and interpretive conversation makers. In this chapter the depths of the forest are released to children's imaginations of how ecological balance can be realised, as illuminated by field notes made during the events of the Living Tree Forest and Wondering Ways.

The authors in this section are all working with projects that challenge the exclusivity and wastefulness of cultural organisations, forming ecologies in practice with local groups and with arts practitioners that reposition relationships between cultural organisations and their human and other-than-human environments.

CHAPTER 12

How Can Artful Learning Help Young People in the Context of the Earth Crisis?

Bridget McKenzie

Introduction

It is a significant challenge to deliver arts learning for young people that addresses the past and future of our planet in ways that are trauma-sensitive, enjoyable, equitable and enabling agency. One challenge is the direct one, which is that the unfolding impacts of the Earth Crisis create real shocks to our places and the systems that our lives and organisations depend upon. More philosophically, for cultural and educational institutions, the tentacular and uncertain impacts of the Earth Crisis challenge the notion of continuity of civilisation that lies behind their founding rationales to collect and preserve into the long-term future (Janes & Sandell, 2019). The intergenerational injustice of this crisis is extreme, calling into question the moral authority of institutions such as museums and universities which are tacitly upholding notions of nationalist competitive growth on a stable, ever-giving planet. In this chapter, I explore

B. McKenzie (✉)
Climate Museum UK, Norwich, UK
e-mail: climatemuseumuk@gmail.com

M. Matthews (ed.), *Ecologies in Practice and Learning: Arts Interventions in the Earth Crisis*, Palgrave Studies in Educational Futures, https://doi.org/10.1007/978-3-031-82338-1_12

the methodological and pedagogical challenges of working in this context, drawing on the experimental practice of creative associates in the Climate Museum UK (CMUK) in supporting young people's responses to the Earth Crisis.

The research summarised in this chapter emerges from questions I have been discussing with my CMUK colleagues, as we reflect on our practice in monthly shared learning meetings.

What follows outlines some examples of our work in different settings with children and young people, with reflections on how best to design artful experiences that create a sense of agency.

The Earth Crisis Changes Everything

Earth systems have been disrupted by centuries of extractive colonialism and the recent acceleration of fossil fuel consumption, ecosystem destruction and pollution. The Stockholm Resilience Centre has quantified the processes, including climate and biodiversity, that regulate the stability and resilience of the Earth (Richardson et al., 2023). Six of these nine boundaries have been significantly breached, with a seventh on the cusp, limiting humanity's potential to thrive for generations to come. Their research has also explored the interaction between the boundaries, creating 'tipping points', exacerbating stresses across all Earth systems, local ecosystems and human systems. In particular, they suggest that climate change is the biggest threat to biosphere integrity, and in turn, damaging biodiverse ecosystems—through acts of ecocide—increases greenhouse gases and worsens climate impacts. Anthropogenic warming can be tracked longer into the past than is assumed, and the worst catastrophes often imagined to come in some indeterminate future are happening now. International climate negotiations are based on the goal of stabilising warming at 1.5C above background temperatures by 2100. However, a third of days in 2023 exceeded 1.5C of global warming (Dale et al., 2023) and 2024 was the first year to exceed 1.5C of global warming, at 1.6C above pre-industrial levels (Copernicus 2025). Impacts are now negatively affecting all areas of the planet, at all latitudes and at all levels of advantage.

While we should not think of the Earth Crisis only as a future issue, the youngest people will have more years of their lives affected by its impacts, and as they develop more awareness of what is unfolding, many are indicating a range of symptoms of 'future anxiety' in anticipation of these impacts. While some might not be articulating this, it is a reality in their lives, and most will be affected by others being conscious of it, including parents, and also by the impact of environmental harms on their bodies. A recent UK study shows that 49% of young people feel anxiety about the future, and 59% agree it feels frightening for their generation (Censuswide, 2022). When the Earth Crisis is overtly in the frame, neutrality in arts and learning is impossible, partly because decarbonisation and political violence are entangled, locked into the historic roots and continuing power of colonialism, and partly because, despite the intergenerational injustice falling on young people, they are the most disenfranchised age-related group in society.

The Earth Crisis is not simply a topic alongside many other subjects. The Earth is the miraculous cosmic body that offers a life-sustaining biosphere, and it is, or should be, the ground of all our learning in spirituality, morality, theory and practices of living. Awareness of the catastrophic disruption to its life-sustaining systems now has to layer on top of comprehensive Earth-centred learning. The outcomes of any learning programme would be to ensure that all people can mitigate harms to living systems through resilience and resistance, and live as well as possible in this disrupted context.

Being Possitopian

The primary principle of CMUK is to be 'possitopian' in thinking about the future (McKenzie, 2020a, b). This is my own neologism for thinking that expands the cone of possible futures and draws on geophysical realities and scientific data, while also harnessing the full powers of the imagination. Being effectively possitopian requires a practice of anticipating futures much more frequently, in many more situations and permutations, and involving a greater diversity of people than is normal in society. Managing the risks of the Earth Crisis is not about working out the best response to the most likely outcome and giving people simple takeaways to carry out in response. It is about determining the better range of responses to the full distribution of possible outcomes and empowering people to discuss and prefigure some of these pathways and to experiment with putting them into effect. In our possitopian activities, we aim to help participants imagine future scenarios which are potentially either more catastrophic or more safe than they might allow themselves to think, drawing on experiences or evidence from the past, and weaving these together about real contexts.

For example, in a workshop for families that we facilitated as part of We Breathe Together, a day on air pollution at the Horniman Museum in

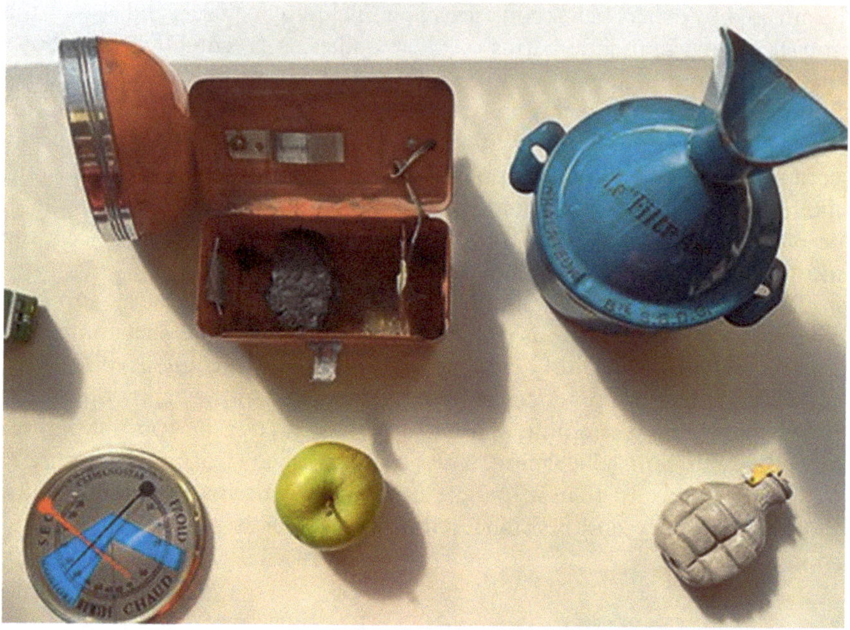

Fig. 12.1 A few of the handling objects used at the We Breathe Together day at the Horniman Museum

London in October 2022, we aimed to expand thinking on how air quality might become worse in future. We invited participants to find and map associations around objects from my handling collection, which I have gathered to illuminate aspects of the Earth Crisis. These included a tea towel decorated with an image of beef cuts, a newspaper from the day of an extreme heatwave, an orangutan cuddly toy, a fruit sweet tin printed with Adam and Eve in paradise, a vintage respiratory inhaler, a coal miner's torch, a piece of coal, a weather monitoring device, a seed bomb and an apple. See a selection in Fig. 12.1.

A well-known factor that worsens air quality is emissions from diesel and petrol vehicles. A lesser-known factor that we introduced is intensive animal agriculture, where particulates from soil disturbance and chemicals from manure and fertilisers used to grow animal feed interact with industrial and transport emissions. In addition to respiratory conditions, there are other negative health effects, such as reduced antibiotic resistance in people consuming intensively farmed meat. Participants were invited to draw and label links across these and many other factors between the objects we had on display. We intended to draw out conversations to grow confidence to speak about provisional and complicated ideas, while also triggering interest in the impacts of industrial production and long-distance transportation on the climate and ecosystems (Fig. 12.1).

We then invited participants to visualise the safe, clean and green future they would work for, and to discuss these possibilities with their family members, while making collages on large scrolls of wallpaper printed with trees. These scrolls were then hung in the museum's conservatory to create a forest of future hopes. Enabling participants to see a situation through different lenses, such as past and future, or visualising the worst and the best outcomes, is a helpful approach to being possitopian (Fig. 12.2).

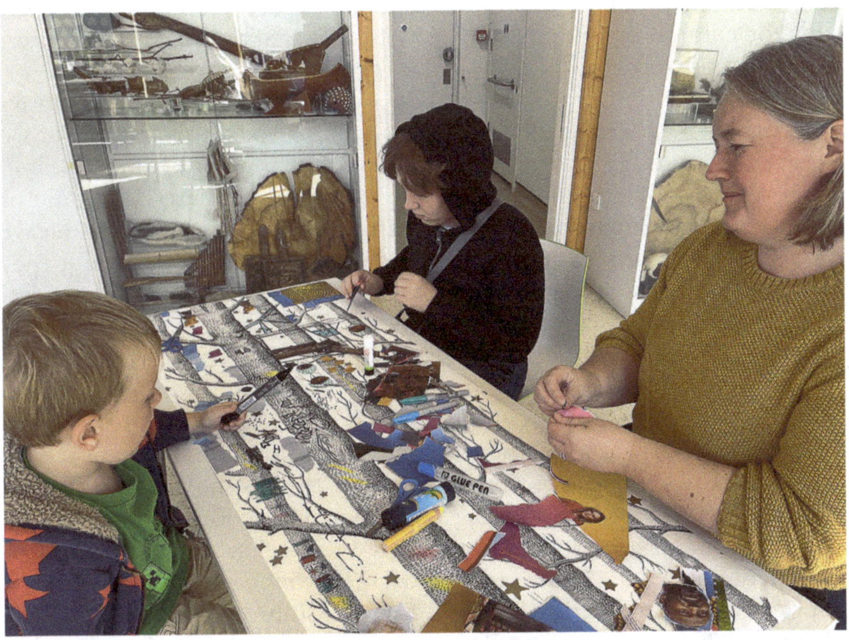

Fig. 12.2 Participants in the We Breathe Together day at the Horniman Museum, making collages

Designing Learning Experiences for Activation

As a distributed and experimental museum, Climate Museum UK is a collective of creative practitioners carrying out activities in our local communities or online. We call these activities 'activations'. Where the default mode of museums and galleries is to collect, show and interpret objects in exhibitions, ours is to engage people in activations. We encourage participants to go beyond simple domestic actions towards empowering themselves to transform their communities, whether in local places, cultural groups, businesses or sectors. We aim to design projects that effectively address the Earth Crisis by therapeutically supporting individuals while also maximising the network effect of social activations for systemic change. We need to help each other in 'staying with the trouble' (Haraway, 2016) and in making trouble to catalyse change. However, this in itself is a troubling challenge for educators wherever the education system is geared towards training people for a growth-based economy and, relatedly, for cultural practitioners wherever culture is geared towards entertainment and commodity.

In the activations that I lead on, I am exploring the most effective ways to design workshops, immersive spaces or art projects which can move participants through a well-held process of sensing, thinking, feeling, imagining possibilities and taking effective action. If there is no time or chance to take such action within an activation, participants should leave with a sense of purpose and clarity about what they can do. Ideally, they should be supported to apply and practise skills for effective actions in real and appropriate contexts such as streets, gardens or school buildings.

Short one-off activities are not generally adequate for such a supported transformation. Quick learning downloads—aiming to be accessible or to fit within the constraints of cultural visits or short lessons—can leave people confused and resistant if they do not tap their intuition or connect with what they already know. A synthesis of 430 studies shows that the least effective method for encouraging environmental action is informational education, and the most effective is 'social contagion' or social comparison (Bergquist et al., 2023). In particular, downloads that start with the impacts of the Earth Crisis can instil fear and shut down the imagination (Turnbull, 2012). Exploring risks and impacts is a vital part of an engaging experience but can only work after trust, safety and modes of expression have been established.

However, typically, providers of cultural learning are not funded or invited to engage with learners or community groups in enough depth or with enough time to establish trust when dealing with such sensitive and complex matters. Educators in formal or informal settings often feel a lack of expertise and support to deal with young people's needs in facing the Earth Crisis. These two points were some of the findings from our 'Listening Project' in which we held focus groups to ask: 'How can culture and creative learning respond to young people's concerns about the Earth Crisis?' (Leach McDonald & McKenzie, 2021).

Ideally, cultural, environmental and educational organisations would collaborate to enable substantial, cross-curricular and long-term programmes to achieve the necessary transformations for young people. In our evaluations of projects in CMUK, we have reflected that such programmes might have some of the following features. First, effective activations are scaffolded so that as the experience progresses, there is an increasing integration of feelings, thoughts and actions. Second, there is a careful mixing of knowledge elicited both from participants' acquired expertise and from scientific research to enable the generation of authentic ideas relevant to real or viable situations. Third, there are plenty of chances for all involved to reflect with each other, going back over things to go forward, and there are regular opportunities to empathise with others, including other biological and cultural perspectives. Fourth, a programme would end by identifying ways that participants can become active agents of a regenerative, responsible system—whether in their own lives or places or on a larger scale—to give them hope for a liveable future.

Designing a satisfying integrated experience is challenging because there is rarely enough time to cognise the complex causes, the 'unthinkable' impacts and the many possible solutions to the Earth Crisis. People need well-held invitations to cycle back to their initial framings and to keep extending from them, opening out to other perspectives. We have observed that artful activities—both artfully designed and using arts-based methods—enhance participants' enjoyment, social interactions and intuition to accelerate the integration of the process. Artful activities can suspend judgment and allow for more rapid acceptance of multiple possibilities. However, if these activities are framed as self-expression or as the practice of aesthetic skills around a loose theme of 'nature' or 'climate' without space for informational enquiry, this expansion of understanding is much less likely. What might be intended as therapeutic will be less effective as such because it has not opened avenues for agency and action.

There are additional challenges in designing experiences with explicit therapeutic purposes for people in states of distress or generational trauma, for example, after experiences of displacement from war or extreme weather. In these cases, it is even more important to be artful in designing the experiences, co-developing with respect and equality, paying attention to pacing, reinforcing or cycling back, explaining processes and offering choices of activity. People with multiple challenges are most likely to feel a lack of agency and to feel overwhelmed by the systemic changes that would address social and environmental problems in multi-solving ways. I try to avoid making optimistic statements or encouraging participants to make pledges for action they might take, unless we can offer routes towards collective efficacy.

Artful Design of Activations for Children

Artful learning is particularly important in designing 'activations' for children and young people. However, there is a paradox here. The younger the person, the less society should expect that they will have practical capabilities or emotional capacities for taking effective action in the face of the Earth Crisis. Yet, the younger the person, the more that they need hands-on and exploratory forms of learning, showing them in concrete ways how actions can have positive outcomes for nature or other people.

An artful approach can be very helpful in this respect because we can draw on several increasingly dynamic strategies. These include using body and voice in improvisatory and energising ways; designing experiences with a sense of resolution; envisaging a simulation in which environmental changes can be imagined; or activities happening in places that offer real potential to apply action ideas directly, such as forest gardens, kitchens, wetland restoration projects or a polluted road.

When we think of the Earth Crisis, top of mind are the most dramatic impacts such as extreme weather, food shortages and displacement of communities. However, young people have lived all their short lives with the slow burn of a degenerative and unhealthy system, which is also part of the Earth Crisis. They are increasingly separated from nature, which includes separation from thriving biodiversity, natural light and darkness, food at its sources, freedoms of expression and play, and chances for outdoor and physical activity. These forms of separation lead to deficiencies affecting physical and mental health, curiosity and knowledge about the diversity of life, and agency to effect change in their surroundings. (McKenzie, 2023).

Overcoming these forms of separation through arts learning is about coming to know relationships between oneself, other humans and other living beings, and then moving towards a sense of re-inhabitation of the biosphere we depend upon. Substantial arts-led enrichment offers potential for learning that shifts learners beyond responses arising from short-term, near-at-hand thinking, to make personal sense of what seems overwhelming, distant or impossible, and to feel more motivated to take action.

As well as the basic forms of separation listed above, young people are disconnected from many levers of agency to improve the systems that affect them, such as the food system and their political representation. Arts-led learning, at least, has the potential to help young people see what is wrong with these systems and to consider what would work better. If effectively designed to respond to the wider Earth Crisis, such programmes could enable them, for example, to run enterprises with ecological innovations, to form activist movements or to restore landscapes. Such programmes would be designed to develop capacities so that, when ready, young people can engage with critical scenarios to imagine disrupted futures, collaborate with others of different generations and pursue actions with tangible benefits to their environments.

However, there is very little recognition of the need and potential for redesign towards these outcomes amongst those who deliver, manage, fund and evaluate arts learning. The report 'Arts in Schools: Foundations for the Future' (Bacon & Tambling, 2023) refers to this being a time of crisis and anxiety but makes no explicit mention of environmental changes and future threats that are strongly affecting young people and no mention of the psychological effects of entanglement in an extractive and exploitative economic system and the encompassing nature of the impacts of global warming, pollution and biodiversity loss.

Learning from Deep Listening

It is vital to listen to and involve young people in enquiries about how they perceive the critical conditions for their lives and the possibilities and challenges for arts learning. This means listening to what they sense, think, feel and want to do. This includes listening out for what they are unable to say and detecting the ways they have been influenced by adults around them, or by advertising or social media discourses.

One cannot assume that all young people will be 'eco-warriors' or climate-conscious, as the generational stereotypes common in the media have increasingly suggested since Greta Thunberg began her School Strike for Climate in 2018. Against this stereotype, many young people are likely to be strongly exposed to attitudes from their elders in schools, faith groups or at home that are disapproving of political resistance or creative expression. Some young people might be influenced by peers or parents who are persuaded by anti-environmental propaganda, such as denial of climate science. Others might be raised within faith-based communities, promoting beliefs that the apocalypse is inevitable and divinely ordained.

We (CMUK) were commissioned by A New Direction to carry out a strand of research for their Listening Project (Leach McDonald & McKenzie, 2021). While the other four enquiries focused on COVID-19, our strand focused on the Earth Crisis as a whole. Our methodology involved artists trained in 'Deep Listening', which is a practice of suspending self-oriented, reactive thinking and opening one's awareness to the unknown. We asked these artists to use sonic and graphic notation to capture what might not be heard by us as we facilitated the discussions. Our headline findings, from listening to young people and adults working with them, included the following points. First, we at CMUK argue that all arts learning programmes must include recognition of the Earth Crisis, and any programme focused on this should be carefully designed for in-depth exploration of its linked dimensions. Second, young people should not be pigeonholed as eco-activists or as suffering eco-anxiety, as all have different attitudes and capacities, and that enabling a meaningful youth voice requires profound awareness of these different needs. Third, cultural educators need trauma-sensitive training to work with young people in the most effective ways when looking squarely at the Earth Crisis.

Most significantly, young people told us how, contrary to assumptions, they do not know enough about the Earth Crisis to feel confident speaking about it so that they can influence change. In particular, they feel they know much less than they need to about how to live in times of collapse and how to build an alternative regenerative system. Society's perceptions of young people as confident and knowledgeable 'eco-warriors' come from media discourses that shine a light on remarkable individuals assumed to have close identity with the majority. An Office of National Statistics study shows that 33% of 16- to 29-year-olds in the UK are very worried about climate change, more than other age groups (Ball et al., 2022). However, the study reports that they are also the age group least likely to have made lifestyle changes in response. Partly, this can be explained by a lack of practical agency, for example, by not being a voter or householder.

It also reflects how strongly influenced by peers, advertising and consumer culture young people can be. An already-cited synthesis study (Bergquist et al., 2023) tells us that 'social contagion' or comparison is the most effective method for behaviour change on climate. Conversely, social contagion can be a big barrier to change when it is propagating the enormous tide of environmental misinformation and consumerist peer pressure (McKenzie, 2024). If young people feel embarrassed to speak about environmental concerns because their peers do not, they will be less motivated to learn about them or build confidence to speak to those who can leverage change. Therefore, there is a strong case for integrating environmental and arts learning, designing experiences to be both enjoyable and serious, where discussing concerns and radical actions is normalised, and the load is lightened with improvisatory fun.

The Imagination of Times Past and Future

In CMUK, we are designing experiences in diverse, experimental ways for affective participation, which include using historical stories and collection objects to stimulate safe but powerful conversations. We use knowledge of the past to inspire the imagining of alternative futures. With increased capacities to imagine backwards and forwards in time, we hypothesise that more young people will be able to take action that prefigures the systemic changes needed.

Our new project 'Make Climate History' uses the Warming Stripes with standardised blue-to-red colours to represent rising global temperatures, developed by Ed Hawkins of the University of Reading (2024). We piloted this project initially in Bromley (London) and Norwich (Norfolk, UK), hoping to offer it to schools and other larger groups across the UK. In February 2024, we participated in the Norwich Science Festival, inviting families to explore how human activities have warmed and changed the climate. Participants created a mural of stripes across the glass wall of the Millennium Library in Norwich, with annotations and drawings of past climate events and future visions, after discovery with library books, a timeline resource and conversations with us (Fig. 12.3).

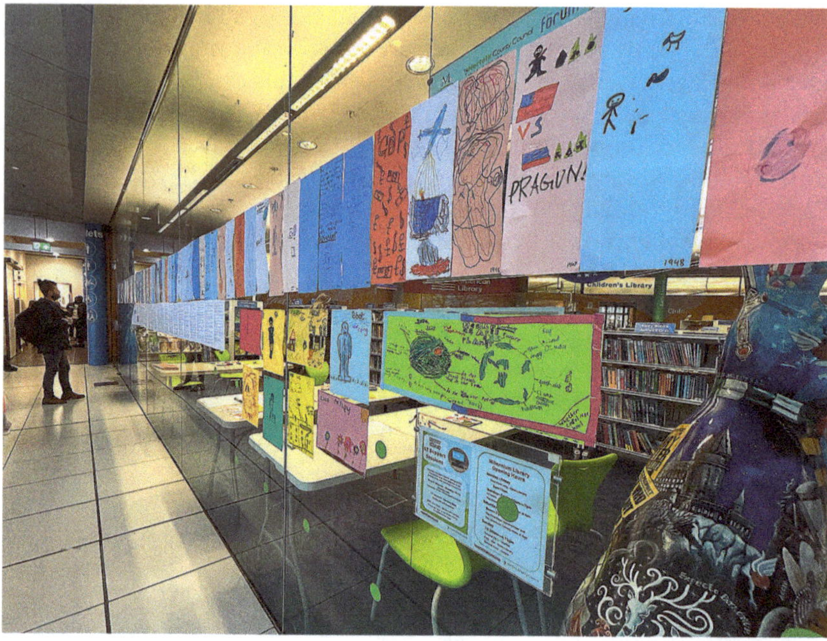

Fig. 12.3 Make Climate History mural installed at Norwich Millennium Library

Norfolk is at a high risk of climate change, with low-lying land and a coastline that is vulnerable to erosion. There have been multiple flood warnings since October 2023: many rural communities are struggling, and farmers are anxious about sodden fields affecting their crops. We heard children echo their parents' experiences and describe them as their own lived experiences. They asked why adults had let global temperatures rise so fast, as 2023 was at 1.5 °C above the pre-industrial global average and ocean temperatures are at extreme highs. They also expressed worries about how pollution affects soil, air and water, pollinators and marine life. They wanted to know how they could help, so we offered some resources about the solutions, and we invited them to create a second artwork to visualise themselves in a chosen future year when their world is greener, safer and kinder.

The third example of our activation projects, which I will share in this chapter, is called Imagine Futures. This was trialled in February 2023 with two secondary school groups in Norwich. A day-long workshop asked the participants: What kind of positive future can you imagine for the human and more-than-human inhabitants in our city, and how can your creativity help? Thinking of your future careers, what will be the creative and helpful roles in the face of the Earth Crisis? The workshop used a 'live action role-play' format, with six adults playing possible roles of the future, such as Healthy Lives Weaver and Water Flow Keeper. These characters had travelled from the future and helped students to generate ideas to help wildlife sites threatened by harmful development and climate breakdown. These local sites were Mousehold Heath, threatened with drought and fire, and the Wensum Valley, threatened by a major road development. The young people's ideas went beyond wildlife conservation to tackle root causes in the decisions we make about energy, health, transport, food and work, using visual methods to develop these ideas using paper-based clay, cardboard structures, collage and large-scale drawings. The project outcomes included a website (https://imaginefutures.net/) offering further workshop plans and resources for young people and arts educators to try their own imaginative work exploring their futures of places and careers.

The intended outcomes of all our projects for young people are to broaden their understanding of Earth Crisis issues and eco-social solutions, and to give them experience of creative methods to express their ideas, so that they can develop confidence to engage peers and influencers. With these two outcomes, we have to overcome tension in the design of our activations, between educating in instructive ways about these issues and enabling explorations through open-ended creativity. Participants pool their environmental knowledge, gently guided by adult or near-peer facilitators, to enrich their interpretations of handling objects or their analysis of problems.

As a collective of around 20 artists and environmental educators, we meet monthly to share our different methods of engagement in supporting discovery of complex scientific and political ideas. With limited time, we need participants to quickly grasp key pieces of information; if these are delivered with too much speed or bluntness, they can be bypassed or be too disturbing to process, limiting the imagination. Our goal is to create rich spaces for young participants to voice their questions and experiences in nuanced ways around the many possible impacts, causes and solutions of the crisis, exploring cultural biases, narratives and conundrums.

Conclusion

The first and essential need in designing arts-led experiences for young people is for adults working with them to fully face the Earth Crisis in its severity and complexity. With more honesty between ourselves as more responsible adults, we can respond more sensitively to the concerns, injustices and confusions that young people deal with. Although young people must grow their knowledge of the systemic causes, impacts on lives and many solutions of the Earth Crisis, the risk of overly factual education is that it creates paralysis of imagination and anxiety, or a diversion into apathy. I argue that arts-led experiences can be more effective if they aim for an integration of a range of modes, such as character role-play, expressive speech and collage-making, in order to support young people to be confident, vocal and empowered. Such experiences can be guided by the possitopian principle (McKenzie, 2020a, b), which is informing our emerging practice of imagining multiple futures beyond the binary limitations of dystopia and utopia. If we put artworks and artistic practices at the centre of an educational experience, balanced with awareness of the Earth Crisis in its wider context, we can create opportunities for the diverse perspectives of participants in activations to combine together so as to generate new depths of expression for 'staying with the trouble' and fresh ideas for making good trouble.

Competing Interests (4) The author has no conflicts of interest to declare that are relevant to the content of this chapter.

References

Bacon, S., & Tambling, P. (2023). *The arts in schools: Foundations for the future.* Calouste Gulbenkian Foundation.

Ball, C., Edser, K., & Windsor-Shellard, B. (2022). *Worries about climate change, Great Britain.* Office of National Statistics.

Bergquist, M., Goldberg, M.H., van der Linden, S., & Thiel, M. et al. (2023), Field interventions for climate change mitigation behaviors: A second-order meta-analysis. In *Proceedings of the National Academy of Sciences (PNAS).*

Censuswide. (2022). Class of Covid campaign report. *The Princes Trust.* Retrieved January 6, 2024, from https://documents.princes-trust.org.uk/Document_TheClassofCovid_Report.pdf

Dale, R., McGrath, M., Poynting, M., & Tauschinski, J. (2023). World breaches key 1.5C warming mark for record number of days. *BBC News.* Retrieved January 3, 2024, from https://www.bbc.co.uk/news/science-environment-66857354

Haraway, D. J. (2016). *Staying with the trouble: Making kin in the Chthulucene*. Duke University Press.

Hawkins, E. (2024). *Show your stripes*. University of Reading. Retrieved July 8, 2024, from https://showyourstripes.info

Janes, R., & Sandell, R. (2019). Editors' introduction: Posterity has arrived. In R. Janes & R. Sandell (Eds.), *Museum activism*. Routledge.

Leach McDonald, R., & McKenzie, B. (2021). How can culture and creative learning respond to young people's concerns about the Earth crisis? *Climate Museum UK for a New Direction*. Retrieved May 15, 2024, from https://www.anewdirection.org.uk/research/listening-projects

McKenzie, B. (2024). Climate misinformation and pro-environmental messaging. *Medium blog*. Retrieved May 15, 2024, from https://bridgetmck.medium.com/climate-misinformation-pro-environmental-messaging-2fba42437638

McKenzie, B. (2020a). Collecting everyday ecocide. *Medium blog*. Retrieved May 15, 2024, from https://medium.com/stories-of-extraction/collecting-everyday-ecocide-cbf6d8f8b5a3

McKenzie, B. (2020b). Explaining Possitopia. *Medium blog*. Retrieved May 15, 2024, from https://bridgetmck.medium.com/explaining-possitopia-96775ad78823

McKenzie, B. (2023). Horizon scan of children and young people, and the cultural sector in the context of the Earth crisis. *Climate Museum UK*. Retrieved May 15, 2024, from https://tinyurl.com/bdzzehaz

Richardson, J., Steffen, W., Lucht, W., Bendtsen, J., & Cornell, S. F. (2023). Earth beyond six of nine planetary boundaries. *Science Advances, 9*(37). https://doi.org/10.1126/sciadv.adh2458

Turnbull, G. (2012). *Trauma: From Lockerbie to 7/7: How trauma affects our minds and how we fight back*. Transworld.

CHAPTER 13

Residents: Re-wilding the Gallery

Trish Scott

Introduction

Years ago, before retraining in the arts, I worked in public policy including, for a two-year period, running the private office of the UK Environment Minister, whose primary responsibility was climate change. I lived and breathed language around the planetary crisis on a daily basis. Upon moving into the cultural sector, where I went on to lead engagement work in Exhibition and Learning teams at major UK arts institutions, I was initially struck by what felt like an absence of everyday discourse around ecological concerns. Sustainability was largely treated as an office management or estates issue connected to reducing the carbon footprint of operations. The climate crisis has since risen up in many curatorial agendas as a programmatic urgency (Buck, 2023). However, it is only more recently that arts organisations are starting to articulate a more complex self-understanding of how they themselves are implicated in perpetuating extractive and exploitative processes not just operationally but via the formulation and enactment of cultural values and beliefs (Foster, 2022).

T. Scott (✉)
King's College, University of London, London, UK
e-mail: trish.scott@kcl.ac.uk

M. Matthews (ed.), *Ecologies in Practice and Learning: Arts Interventions in the Earth Crisis*, Palgrave Studies in Educational Futures, https://doi.org/10.1007/978-3-031-82338-1_13

In this text, I briefly explore the implications of unsustainable cultural practices on gallery communities and, through a case study of a particular project, *Residents*, at Goldsmiths Centre for Contemporary Art (the CCA), offer an alternative institutional operating model for hosting the arts rooted in ecological, as opposed to colonial capitalist, logic.

Residents is an experiment at the intersection of education and curation focused on opening-up and re-modelling gallery space and systems, breaking down institutional boundaries to 're-wild' the relationship an arts institution can have with its constituting communities (Yung et al., 2015; Foster, 2022).

From an ecological perspective, re-wilding is concerned with protecting and restoring biodiversity. It is often understood as a self-healing approach to conservation that embraces uncertainty, openness, and the complexity of human-nature interactions. Its emphasis is on 'withdrawing anthropogenic interventions' (Massenberg et al., 2023, p. 45) and 're-building ecosystems that have previously been modified by human disturbance' (IUCN, 2021).

In cultural terms, re-wilding can be understood as dismantling conventional and inherited structures and ways of working. It involves re-orienting care, attention, and agency towards 'under-represented, absent, non-standard or uncommon values and relationships' (Yung et al., 2015, p. 24). In its most progressive form, it is closely tied to decolonisation and efforts to restructure the cultural landscape to decentre whiteness and foreground the agendas and needs of those who, historically, have experienced discrimination, inequality, and reduced opportunities due to race and intersecting protected characteristics.

As I will explore in this chapter, *Residents* is an experiment in cultural re-wilding; an exercise in re-balancing a gallery's ecology to become more inclusive and sustainable whilst also addressing calls for 'collectivity against structural and institutional racism in the cultural sector' (Monserrat et al., 2020). *Residents* adopts ecological modelling (Foster, 2022) to re-imagine the CCA as a pluriverse (Escobar, 2018), a non-hierarchical space where multiple worldviews can co-exist.

Before introducing *Residents* and exploring these concepts in more detail, I want to briefly outline how I am defining a growth model of cultural production and how this manifests, particularly for gallery educators working with communities.

Unsustainable Cultural Production

As Foster observes, 'successful' arts leadership is often equated with adherence to a capitalist business model, measured in terms of factors such as 'budget size, growth, fiscal discipline, and capital accumulation' (2022, p. 19). When translated into curatorial practice, this operational formulation places a value on doing 'more'. In what is described as a 'crisis in curating', curators are continually pressured to produce more spectacular shows, at a bigger scale and faster pace. As Krasny et al. point out, this runs the risk of reducing curating to 'glossy selectionism, taste-making, and the overall commodification and commercialisation of culture for consumption and culture as spectacle' (2023, p. 2). Extractive and exploitative ways of working essentially favour 'show and tell' curatorial formats and perpetuate a culture of heroic individualism where 'star' individuals (e.g. artists, curators, leaders) are celebrated, and other agents in a production chain (both human and non-human) become depleted, devalued, and burnt-out. Under a capitalist growth model, the art world has normalised the 'exploitation of care' which, in curatorial terms, is predominantly delivered by 'underpaid and overworked' female staff and freelancers (Krasny et al., 2021, p. 20).

Within this working paradigm, gallery educators often find themselves in an even more challenging position than their curatorial counterparts because of further symbolic divisions and differential forms of prestige that are accorded to different activities within the cultural matrix. As Reckitt observes, 'within the cultural field, high status typically accrues to activities linked to artworks and artists (especially if they are critically or commercially validated)'—i.e. exhibition-making work conventionally done by curators, 'while low status adheres to tasks involving 'uninitiated' groups, such as children and non-art world audiences' i.e. the work conventionally done by educators (2023, p. 173). Because curators typically align with the 'self' of the artist, and educators with the 'other' of the visitor, educational work is accorded lower status and value (Allen, 2009). Status differentials are further compounded by the fact that those tasked with making art institutions more 'relevant, inclusive and accessible', i.e. the educators, tend to be people who are the 'most impacted by injustice and underfunded' themselves (Schweiker, 2023).

Status differences have practical ramifications. In spatial terms, curatorial work is usually highly visible and placed at the front and centre of institutions. Educational work is often treated as secondary or supportive and conventionally assigned to more peripheral spaces. In terms of agenda setting, educators are frequently called on to work flexibly, responsively, and to very short production deadlines, reacting to directives set by curatorial colleagues and other managers with limited broader strategic agency.

Organisational design, rooted in business modelling, compounds these power differentials by reinforcing siloed working practices which can block co-operation (Foster, 2022; Rowson & Villeneuve, 2017). For example, in a functional division of labour, where staff are divided into separate learning and exhibition teams and differently located, it can be hard to collaborate, even if the will is there.

Unsustainable Community Engagement

In a culture of exploitation and extraction, internal hierarchies and tensions are replicated in relationships with audiences, particularly communities that make up the local ecosystem. For example, if a local resident is invited to participate in a gallery education programme 'out the back' or 'after hours', this positioning carries with it a certain message around institutional value. Whilst often accounted for in relation to language rooted in benevolence, i.e. 'you're in this space at this time because you'll have the place to yourself and won't be interrupted', it's hard not to see this as a reinforcement of status norms around who gets 'seen' and what gets prioritised during prime time in gallery environments.

In recent years, the importance of re-centring practice on audiences has gained increasing traction across different related fields. From Simon's work on the art of relevance and inclusive organisational change (2016) to Walmsley's plea for more 'sustained, authentic, relational and effective approaches to audience engagement' (2019, p. 2) to Janes and Sandell's work around museum activism, the necessity of building strong relationships with local communities is clear (2019).

Whilst calls for community-centred approaches have led to an increased focus on audience engagement and partnership working, Lynch's research demonstrates that relations between host organisations and community groups aren't always partnerships on an equal footing. Lynch notes that organisations often manipulate an 'illusion' of consensus and adopt a narrative of generosity, positioning communities as 'beneficiaries', rather than treating partners as 'active agents' (2011, 2020).

Not only do narratives couched in generosity perpetuate deficit model thinking, rooted in colonial capitalism, by positioning larger, often white, organisations as benevolent providers of something structurally marginalised communities are 'missing', but the question of who benefits can, in reality, be the other way round. It is not uncommon for dynamics around engagement to be superficial and self-interested, with larger organisations partnering with smaller community groups or individuals primarily to enhance their own status and standing, for example, in order to appear more diverse and attractive to potential funders. As Montserrat et al. note: 'If there are just enough Black and Brown bodies in the room does your institution and its culture of box ticking breathe a sigh of relief?' (2020).

Re-balancing Power with Communities

In outlining the dynamics which can characterise partnerships with communities and educational work under a capitalist growth model, there are obviously many exceptions to the general picture I have painted. Indeed, much of my work with galleries and museums over the last decade has focused on re-thinking the power balance inherent in the relationship between artists, curators, educators, and audiences, and finding ways for those in a typically subservient position in the cultural production matrix to genuinely exercise authority, troubling boundaries (that characterise more traditional, hierarchical ecologies) between curating and learning, inside and outside, centre, and periphery.

Journeys with 'The Waste Land' (2018) at Turner Contemporary explored what could happen when audiences were given full licence and support to curate a major exhibition. *Spaceship School* (2015) and *Together* (2018–2019) at Tate were alternative versions of art schools based on the premise of children teaching adults. In both these scenarios the usual power dynamics inherent in gallery programming were reversed, with those normally in a 'recipient' role or position of less power invited (and supported) to exercise leadership and set the terms for what might constitute an 'exhibition' in the first instance and a 'learning programme' in the second. The projects existed at the intersection of experimental education and curatorial activism. Neither were about amplifying a pre-configured curatorial offer to broader audiences but were about furthering participatory democracy (Lynch, 2020) through placing value on lived experiences and de-centring normative modes of authority in favour of alternative voices.

I have often come back to consider how modes of activism, which define audience-focused projects of this nature, might be applied more extensively to core gallery operations. For example, how democratic models of practice can extend beyond the realm of what engagement specialists might facilitate at a temporary programmatic level, to influence and transform the very constitution of institutions, with the aim of building long-term, inclusive sustainability.

Goldsmiths CCA: The Context

In 2019, I joined the CCA to lead on engagement at the newly opened gallery. This presented an ideal opportunity to critically consider the role of learning within an arts organisation and re-think conventions of 'engagement' being focused solely on work with external communities (i.e. schools, families, and young people) in favour of seeing it as a holistic practice relevant to broader organisational development and design. In initiating work at the CCA, rather than reproducing siloed thinking, or pre-prescribing relationships and outcomes, I was keen to work in a bespoke and environmentally attuned way.

As a publicly facing gallery on a university campus, the CCA occupies a unique position in Southeast London. It sits at the intersection of different communities, and I was interested in its potential to nurture new forms of knowledge exchange, collaboration, and connectivity, at the juncture of art, academia, and lived experience.

Residents, as a nascent idea, emerged soon after my arrival. To build an understanding of the gallery's context, local ecosystem, and neighbourhood needs, I undertook an environmental scanning exercise (Foster, 2022) in collaboration with an interdisciplinary group of students. This work evolved into a co-devised proposal to host a small number of community groups at the CCA to test out what different constituents might want from a new visual arts centre, and what they might need to feel 'at home' in the gallery space itself.

Owing to the pandemic, this work did not progress as planned. However, a further period of organisational self-reflexivity in the wake of George Floyd's death, Black Lives Matter conversations, and the sectoral crisis of representation that ensued within the art world led both to the development of an anti-racist action plan for the gallery, and the inception and realisation of *Residents*, a bigger cross-gallery initiative intended to open-up pathways to belonging and exchange.

Setting up *Residents*

Residents, launched in 2021, is a user-led programme and relational constellation which involved the transformation of the CCA's ground floor gallery into a multi-functional site able to serve the needs, and host the work, of up to ten rotating local communities of practice simultaneously and over the long term (Wenger, 1999). Rather than engagement being in response to exhibitions or a component of artist-led projects, *Residents* put community interests and needs first and started with an open offer for local people to take up (and transform) gallery space, funding, and other resources at the CCA, based on the premise of 'usership' (Wright, 2013). 'What would you find useful from a relationship with us and how can we support you to do your work?' community groups were each asked as they become involved.

At its heart, *Residents* was predicated on a complete re-think of gallery norms and drivers both spatially, relationally and programmatically. It was set up to challenge conventions about 'expert culture', stemming from colonial capitalism, and to support the needs of those most affected by extractive and exploitative traditions. *Residents* was underpinned by a re-direction of curatorial care, space and resources towards artists, groups and practices traditionally marginalised by the cultural sector, and who were living, studying or working locally.

In setting up *Residents,* conversations explored how to add value to what was already happening in the locale in order to amplify under-represented cultural forms, practices, and networks deemed locally relevant that needed more support.

Residents began with a small number of groups (CCA, 2021). Several groups became involved through invitation, in order to deepen and sustain existing links and connections with the CCA. These early joiners helped shape parameters for the growing programme. Others came onboard through nomination or in response to an open call issued to students. Groups ranged in scope and scale from a neighbouring school, to a disability-arts collective, to an intergenerational community of interest that had formed around the rare groove music scene, to a collective of students, alumni, and educators from three different universities all engaged in building creative learning environments based upon the values of equity and care. The initiation and design of *Residents* was developed collaboratively at an organisational level and supported through reciprocal conversations with local advisors of colour. As 'critical friends', advisors were resourced to work in tandem with the CCA's core team, who, at the time were predominantly white and European.

Cultural Re-wilding

In leading with relationships and working out how to support diverse local voices and practices to become an active and sustained part of the CCA community, centred on groups' interests, it was necessary to take an open and organic approach towards gallery practice and development. In this way, *Residents* grew from ecological thinking as opposed to classic business modelling. In structural terms, *Residents* embodies Foster's conception of ecological leadership (2022). As Foster notes, when an arts organisation starts to think of itself as an ecosystem, i.e. a complex and interconnected web of people, things, spaces, ideas and processes, situated within a constantly changing environment, 'its perspective about organisational design and practice will shift from vertical and linear to horizontal and omnidirectional' (2022, p. 39). This involves moving away from individualistic, competitive organisational thinking ('I' oriented thinking), and towards an understanding centred on nurturing the health and sustainability of a bigger whole ('we' oriented thinking). In understanding the CCA as part of a larger, dynamic whole *Residents* takes a systems-oriented approach, emphasising 'interconnectedness and interdependence' (Jung & Love, 2017, p. xiii).

As an exercise in cultural re-wilding *Residents* was conceived as open-ended, interdisciplinary experiment initiated to change the balance of the CCA as a community and network. In disengaging from 'artificial, inherited structures' and systems of domestication *Residents* focused on engaging with diverse local voices to 'renegotiate relations together anew' (Yung et al., 2015, p. 24), striving for a more sustainable formulation for how a gallery might be composed, conceived of, and used.

Towards a Pluriverse

Spatial re-design was a key component of *Residents.* In order to meet the needs of different resident groups and be open and permeable to different forms of knowledge and ways of doing things, it was necessary to transform the CCA's ground floor from a gallery into a flexible, multi-functional space. To achieve this, the first resident groups spent summer 2021 appointing and then working with design collective *Access, Power &*

Visibility (APV) to co-design a new version of the CCA's ground floor gallery via a series of workshops.

The positioning of *Residents* at the front and centre of the CCA, rather than 'out the back' was a deliberate move to invert both spatial conventions and status differentials typically associated with different people and practices within a gallery production matrix. Earmarking the CCA's front gallery for *Residents* was about attaching symbolic importance to this work and to the communities who would be using it. However, at the same time, it was important for residents not to inadvertently become a spectacle, or for anyone to feel that they had been invited in to 'perform' their community status. It was fundamental that groups had agency and control over their own spatial thresholds, and could decide themselves how much visibility they wanted vis a vis the CCA team and other visitors.

APV's final design (Fig. 13.1) rooted in the social model of disability (APV 2023), focused on accessibility, hospitality, and welcome as foremost concerns, with an emphasis on supporting the development of a 'living, breathing space' that emphasised people, practice, process, comfort, care, and connection (*Residents* Design Workshop, August 2022). Defining features were a carpet, changeable coloured lighting, curtains that could de-mark different zones to facilitate differing levels of privacy and intimacy, and custom-built modular storage which enabled Resident groups to quickly change the set up and feel of the space, depending on who was in and what was happening from day to day.

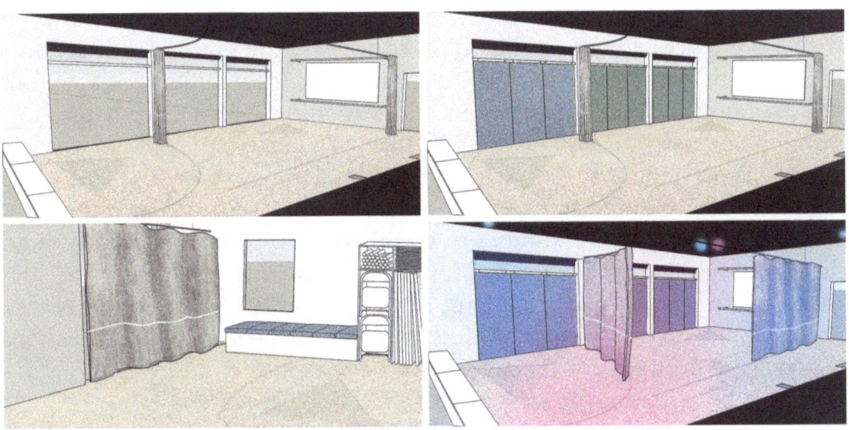

Fig. 13.1 CCA *Residents* Design Document (APV, 2021)

Conversations around the connected nature of different senses, the inter-relationship of mind and body, and the importance of supporting non-verbal and non-visual modes of knowing led to a sensorially oriented design, which was particularly 'inclusive to touch' and enabled a 'never-ending-ness of attention and focus of play' (Entelechy, n.d.). Certain features of the design, such as a carpeted floor, challenged traditional gallery aesthetics, but were key to creating a 'living room' both in the sense of a 'living, breathing space' and in the sense of helping groups to feel 'at home'. The carpet, for example, afforded new, more relaxed ways of interacting, premised on lying down or stretching out, inviting new attention to bodily practice. Groups started co-habiting space at the CCA in autumn 2021, with each Resident community using the gallery at different times throughout the week, and in different ways. Rather than using the space consecutively, i.e. for a month each over the course of the year, a defining feature of *Residents* was that groups had a regular and sustained presence at the CCA over the long term, to enable the work of Resident communities to become part of the organisation's 'everyday'.

Depending on the day, the CCA's ground floor shifted from being a performance venue, to a classroom, to an events space, to a rehearsal room, to a meditation space, to a film studio, to a temporary exhibition space to a screening room, in accordance with each group's needs and agendas (Fig. 13.2). This enabled the emergence of a new ecology supporting multiple-use values and inclusivity.

For some groups, such as *BlackGold*, a collective of postgraduate artists from various nationalities, ethnic and diasporic backgrounds, united in their experience of being 'othered' in education, the space functioned as a sanctuary from the outside world; a space to work on self-care, away from the pressure of oppressive systems and structures and demands for constant productivity (*Residents* Sharing Event 2022).

Others used the space to explore visibility and invert daily norms around who gets 'seen' (and how). *Ambient Jam*, a multi-media dance improvisation group, who have a number of members with complex and profound disabilities, staged 'durational happenings', inviting the public to join them in the act of 'deep hanging out' (Entelechy 2023). *Floor Five*, a multi-disciplinary collective of Goldsmiths graduates, offered workshops led by women of colour to 'make visible the identities and ways of being that have been absent or distorted within canonical representation' (CCA, 2021) and, as part of this, tapped into the CCA's marketing and communication channels to support and amplify their work.

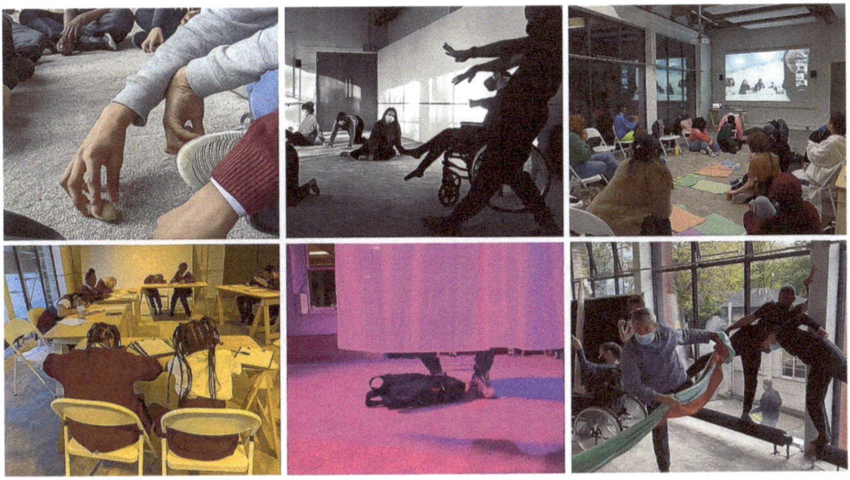

Fig. 13.2 CCA *Residents* Space in Use (From left to right: Kirkpatrick (**e**), Scott (**a** and **d**), Swift (**b** and **f**), Telling (c), 2021)

Essentially, *Residents* came to mean different things to different people, and groups drew on resources offered (e.g. space, funding, materials, marketing, and mentoring) in different ways, to different ends, each retaining autonomy over choices, and the discursive space they occupied.

In sharing one space, but which had multiple manifestations, *Residents* can be understood as a 'pluriverse' or 'a world where many worlds fit' (Escobar, 2018, p. 4). Whilst colonial, capitalist logic emphasises homogeneity and a singular worldview, Escobar explores design as a tool for change. He argues that to design spaces and structures is simultaneously to design 'ways of being' (2018, p. 4). And, in designing for the pluriverse, with a focus on multiplicity and self-sustaining ecologies, diverse perspectives and practices can start to co-exist within a shared space and framework, offering a challenge to any one dominant worldview. In this way, design becomes an activist tool for re-conceptualising and re-building communities.

As a pluriverse, *Residents* enabled people to find a connection with the CCA but for different reasons and in different ways, softening both physical and conceptual barriers with the institution.

But Has Anything *Really* Changed?

Putting best intentions aside, does *Residents really* offer an emancipatory model from the extractive, exploitative, and exhaustive conventions of colonial capitalism, in terms of gallery practice and development, or, at some level, are inherited structures and power differentials inescapable? As Lynch reminds us, "'Invited spaces" in museums, as elsewhere, are ostensibly devices for dialogue, but remain forever permeated with the power effects of difference' (2011, p. 147).

What Lynch's work brings into sharp focus is that even the most genuine attempts at fostering collaborative partnerships with communities can meet with obstacles, often 'inadvertently created by institutions themselves' and 'invisible' to the professionals involved. This is due to power dynamics being deep-rooted and insidious, which, if not recognised can block progress and even be damaging if couched in 'politics of kindness' (2020, p. 5).

So, how successful has *Residents* been in countering inherited dynamics and in shifting organisational practice at the CCA? To what degree has re-wilding been achieved?

There are perhaps several different answers to this question. Certainly, *Residents* has changed the spatial constitution of the CCA. It has effected a re-direction of resources towards local constituents, based on supporting what communities themselves have identified as 'useful' (Lynch, 2020). It has interrupted norms around spectatorship, casting attention on experiential and pedagogical practices (typically marginalised or de-valued in cultural spaces). *Residents* avoids a version of engagement which is siloed, which pre-prescribes outcomes, and which segments people into artificial categories. Instead, it celebrates open-ended, organic interdisciplinarity rooted in collectivity.

I should highlight that my narrative and observations around *Residents* are based on having led on setting up the programme. In its sustained form, the programme is now managed by artist and educator Adrianna Whittingham, who was in residence themselves during the programme's first year, as a member of *Floor Five Collective*, and is now part of the CCA's core team, leading *Residents'* future constellation. Whittingham's appointment is an example of how *Residents* can be understood to be 'growing' into the CCA in institutional terms, overcoming boundaries that typically keep different constituents of a gallery's community (e.g. staff, artists, local residents and visitors) separate and compartmentalised.

However, although *Residents* offers an innovative model for engagement that, structurally, has permeated the CCA's core team, it is also important to note that *Residents* is just one part of a bigger system. The gallery's basement and first floor continue to host exhibitions, and the CCA's public identity is still, also, predominantly exhibition-focused. Adopting a low-key public approach towards *Residents* was a conscious decision to give full discursive space to residents themselves and to ensure that no inappropriate claims of institutional ownership would be made over what groups were doing. However, at the same time, it could be argued that this has inadvertently kept *Residents* hidden, perpetuating an ongoing power differential that, although intended to demonstrate care and safeguard groups' independence, at another level might still be seen as falling prey to dichotomous 'us' and 'them' thinking.

I am conscious of describing *Residents* in effusive terms, but I want to end by noting that setting up *Residents* was not without conflict and involved ongoing negotiations across departments and communities in order to determine priorities at any given moment.

Sound was a common flashpoint. Because of apertures in the fabric of the CCA's building, there is acoustic bleed between floors and spaces. What this means practically is that a *Residents* gathering involving music could lead to sound seeping into exhibition spaces, potentially altering the experience of an artist's work that other visitors had come to see. In an instance like this, what and who should take precedence? In a model emphasising 'usefulness', were residents free to do anything, at any time? Or were there moments when use, from an institutional point of view, became 'misuse' (Wright, 2013, p. 66)?

In a traditional model of gallery practice, community engagement would typically be subservient to exhibitionary activity. However, as Whittingham notes, *Residents* has involved much more 'give and take' (personal communication August 2023). Spectatorship, as a mode of address, still exists at the CCA, but it is not always a given. It is negotiated. This means that different value systems, rather than being unconsciously reproduced, are actively being explored and understood anew, particularly when working through moments of tension.

What *Residents* reveals is that cultural re-wilding is complex. It is not about simply replacing one approach with another but is an involved and complicated exercise in re-balancing care, attention, and agency. It is a journey that involves considering the needs of multiple, inter-dependent constituents who make up an ecosystem at any one time and a journey that the institution and its collaborating communities are still on.

Competing Interests The author has no conflicts of interest to declare that are relevant to the content of this chapter.

References

Allen, F. (2009). Border crossing. In F. Allen (Ed.), *Documents of contemporary art: Education (2011)* (pp. 205–207). Whitechapel Gallery & MIT Press.

APV. (2021). Info. *APV.* Retrieved August 8, 2023, from http://www.accesspowervisibility.com

Buck, L. (2023). Curators in the climate crisis. *The Art Newspaper.* Retrieved January 20, 2024, from https://www.theartnewspaper.com/2023/05/22/curators-in-the-climate-crisis-who-are-the-new-museum-hires-helping-to-turn-their-institutions-green

CCA. (2021). *Residents.* Goldsmiths CCA. Retrieved August 1, 2023, from http://web.archive.org/web/20211016215349/https://goldsmithscca.art/residents/

Entelechy Arts. (n.d.) Ambient jam collective in residence at goldsmiths CCA. *Entelechy Arts.* Retrieved August 8, 2023, from https://entelechyarts.org/projects/ambient-jam-collective-in-residency-at-goldsmiths-cca/

Escobar, A. (2018). *Designs for the pluriverse: Radical interdependence, autonomy, and the making of worlds.* Duke University Press.

Foster, K. (2022). *Arts and cultural leadership: Creating sustainable arts organizations* (2nd ed.). Routledge.

IUCN. (2021). *The benefits and risks of re-wilding.* Retrieved August 15, 2023, from https://www.iucn.org/resources/issues-brief/benefits-and-risks-rewilding

Janes, R. R., & Sandell, R. (2019). *Museum activism.* Routledge.

Jung, Y., & Rowson Love, A. (2017). *Systems thinking in museums: Theory and practice.* Rowman & Littlefield.

Krasny, E., & Perry, L. (2023). *Curating with care.* Routledge.

Krasny, E., Lingg, S., Fritsch, L., Bosold, B., & Hoffmann, V. (Eds.). (2021). *Radicalizing care: Feminist and queer activism in curating.* Sternberg Press.

Lynch, B. (2011). Collaboration, contestation, and creative conflict: On the efficacy of museum community partnerships. In J. Marstine (Ed.), *The Routledge companion to museum ethics; redefining ethics for the twenty-first-century museum* (pp. 146–163). Routledge.

Lynch, B. (2020). Introduction: Neither helpful nor unhelpful—A clear way forward for the useful museum. In A. Chynoweth, B. Lynch, K. Peterson, & S. Smed (Eds.), *Museums and social change: Challenging the unhelpful museum* (pp. 26–57). Routledge.

Massenberg, J. R., Schiller, J., & Schröter-Schlaack, C. (2023). Towards a holistic approach to rewilding in cultural landscapes. *People and Nature, 5*(1), 45–56.

Montserrat, J., Wee, C., Williams Gamaker, M., Ateh, T. et al. (UK Artists4BLM) (2020). We need collectivity against structural and institutional racism in the cultural sector. *Arts Professional.* Retrieved August 1, 2023, from https://www.artsprofessional.co.uk/magazine/article/we-need-collectivity-against-structural-and-institutional-racism-cultural-sector

Reckitt, H. (2023). From coping to curious: Unlearning and reimagining habits of care. In E. Krasny & L. Perry (Eds.), *Curating with care* (pp. 169–196). Routledge.

Rowson Love, A., & Villeneuve, P. (2017). *Visitor centred exhibitions and edu-curation in art museums.* Rowman & Littlefield.

Schweiker, R. (2023). *Hello. Manchester.* The Whitworth.

Simon, N. (2016). *The art of relevance.* Retrieved August 10, 2023, from https://artofrelevance.org/read-online/

Walmsley, B. (2019). *Audience engagement in the performing arts: A critical analysis.* Springer International Publishing.

Wenger, E. (1999). *Communities of practice: Learning, meaning, and identity.* Cambridge University Press.

Wright, S. (2013). Toward a lexicon of usership. *Van Abbemuseum.* Retrieved August 17, 2023, from https://www.arte-util.org/cms/wp-content/uploads/2015/03/Toward-a-lexicon-of-usership.pdf

Yung, H., Peerbaye, S., & Verstappen, M. (2015). Rewilding the arts ecosystem: A discussion paper on multidisciplinarity in the arts in Canada. *CPAF.* Retrieved 15 August, 2023, from https://artisticintelligence.com/resources/Rewilding_the_arts_ecosystem-Multidisciplinarity_in_the_arts_in_Canada-July_2015.pdf

CHAPTER 14

Forest of Imagination and the Living Tree

Penny Hay and Andrew Amondson

Introduction

This chapter focuses on creative research embodied in two recent living installations co-created with Forest of Imagination, an annual participatory contemporary arts and architecture event in Bath, UK. Forest of Imagination is free and open to all, with a creative learning programme delivered by a collaborative team of local organisations in Bath, reimagining familiar spaces, inspiring creativity, and heightening awareness of nature in response to the ecological emergency and the Anthropocene (jagodzinski, 2018).

Forest of Imagination is a creative ecosystem for art and participation. Spaces in an urban context are reinvented to inspire spontaneous play, unleash imagination, heighten our engagement with nature in the city environment, and shine a light on the importance of local and global forests' capacity to inspire creativity in everyone.

P. Hay (✉)
Bath Spa University, Bath, UK
e-mail: p.hay@bathspa.ac.uk

A. Amondson
https://www.andrewamondson.com/, Berlin, Germany

M. Matthews (ed.), *Ecologies in Practice and Learning: Arts Interventions in the Earth Crisis*, Palgrave Studies in Educational Futures, https://doi.org/10.1007/978-3-031-82338-1_14

This long-term research project is an example of civic engagement and innovation extending the partners' civic role through creative, participatory public engagement, creative, cultural partnerships, and widening participation projects with schools and colleges to generate new ideas and approaches with the educational, creative, and cultural sectors. It is a unique collaboration which has developed a vibrant programme to bring a large and diverse audience to engage in Forest of Imagination and its themes of creativity, the environment, and education. Unlike more short-term or prescriptive projects, Forest of Imagination exemplifies an approach to cultural change with creative pedagogy and promotes co-creation and collaboration with a range of diverse partners, including schools and cultural centres. The challenges of bringing uncertainty and unknown outcomes into the co-design process cannot be underestimated. As a research and public engagement project, it invites everyone to have a conversation about the importance of nature connection and collective imagination in a playful immersive environment for all ages.

> Forest of Imagination is a place where everyone can explore their creativity. It offers a re-imagining of a familiar space into a fantastical world to inspire intuitive play, and imaginative thought to heighten a sense of nature in the city. Part landscape, part city, part gallery, the Forest engages everyone in a seriously playful and alternative experience of nature, contemporary art, landscape and architecture. The Forest is a wild place, a metaphorical retreat from the everyday urban world, with delight and darkness, colour and light, re-purposed, natural and intelligent materials, to share the innate creativity of human nature.
>
> Landscape architect Andrew Grant, co-founder, Forest of Imagination (2024)

The creative methodology includes integrated evaluation and multi-modal case studies (film, photography, text, journaling), distilling the key learning from the research embedded in Forest of Imagination's Living Tree Mirror Maze (2022) and Wondering Ways (Amondson, 2023) (Fig. 14.1).

Children's enquiries were documented by the Forest of Imagination team as in-depth case studies addressing the artists' design and intent for Living Tree Mirror Maze and Wondering Ways, and an analysis of the children's ways of enquiring and their interpretation, debate, thoughts, and theories, in collaboration with artists.

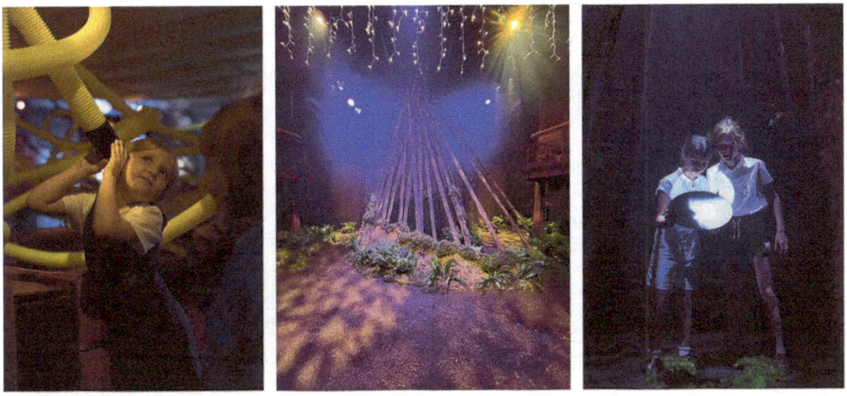

Fig. 14.1 Living Tree Forest Mirror Maze 2022

The following poem, offered in place of a descriptive text for the art installation *Wondering Ways*, highlights one of the guiding principles of Forest of Imagination, that learning should never be prescriptive and that every child should be given space, time, and attention to follow their fascinations.

Wondering Ways

by Andrew Amondson

Where are we
What is the time
How far have we wondered
and who will finish our rhymes? Fig. 14.2)

Where Are We?

What we have to do now is work with nature. We have to see nature as something not to dominate but as our collaborator, our friend, and if we're lucky, the source of our survival. (Heffernan, 2022a, b)

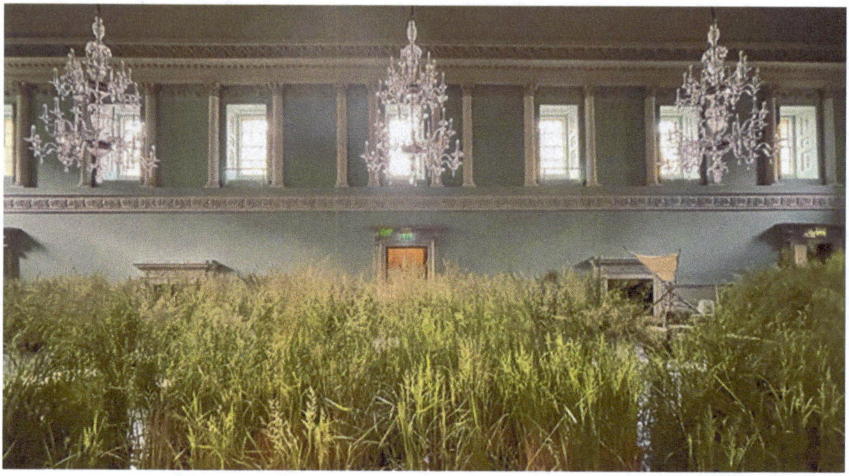

Fig. 14.2 Wondering Ways 2023

Recent studies suggest that climate change education is no longer about avoiding the disastrous effects of climate change but about adapting, and for our survival as a species, this will not be possible unless we fundamentally change our systems of education.

The education system in many countries is broken. Drawing on Sir Ken Robinson's (2021) thinking on how the Industrial Revolution has wrongly shaped our systems of education, creative practices are being eliminated from many curricula. Rote memorisation and standardised testing are replacing creative problem-solving and critical thinking. A love of learning has been co-opted by a market economy, comparing schools and growing children in silos in service of our gross domestic product. The Industrial Revolution set a system of education in motion that is no longer viable and needs a paradigm shift (Robinson & Robinson, 2021).

At the same time, we have drifted further and further away from the nature that nurtures us, standing at the precipice of our prideful separation from it. We no longer remember the names of the plants that heal us in our Western neoliberal civilizations; we have lost the path back to our original purpose, to be good stewards of the earth. We are at a tipping point and are poised to lose everything because we no longer value the very thing that keeps us alive. Indigenous wisdom and custodial practices must be honoured and applied. Although indigenous peoples only make up 5 percent of the global population, they are protecting 80 percent of the world's remaining biodiversity (State of the Environment Report, 2021).

> One of our responsibilities as human people is to find ways to enter into reciprocity with the more-than-human world. We can do it through gratitude, through ceremony, through land stewardship, science, art, and in everyday acts of practical reverence. (Wall Kimmerer, 2015, p. 190)

This understanding is central to the approach in Forest of Imagination, one that places nature connection and climate action at the heart of the work.

What Is the Time?

NOW IS THE TIME. As a society, we need to reimagine the education system to place nature, imagination, the arts, and well-being at the heart. Forest of Imagination champions this change by co-creating and researching experimental sites for pedagogical innovation, placing young people's agency at the centre. It creates an opportunity for children and adults to think differently about the way we see nature, to imagine new possibilities, and hopeful futures. We have been thinking deeply about climate education and how we can respond with hopeful and imaginative solutions to the ecological emergency in education, pedagogical imaginations, thoughts, and practices (Paulsen et al., 2022).

> For ten years, a group of artists and educators have set up a summer camp in the centre of Bath and invited children to come and play with them, creatively exploring nature and humanity's place within it. The invitation changes according to who is involved, where the activity is happening, and what resources are available. One year, a thousand butterflies floated through Bath Abbey, another, a huge white rabbit sat in Queen's Square; during the pandemic, the event went online. In 2023, they transformed the historic Assembly Rooms into a series of installations, exhibitions, and workshop spaces that are free and open to all. The ballroom is a prairie of grasses turning into hay; elsewhere, dead box bushes hang from the ceiling, a beaver's home stands in a silver river, a ceramic moon glows in the dark… it's beautiful, intriguing and thought-provoking. (Matarasso, 2023)

Forest of Imagination invites a conversation about our connectedness to the natural environment and our collective imagination, highlighting the importance of environmental education and environmental stewardship. Forest gives permission for a new way of experiencing nature in a city environment so that the city becomes a place for new approaches to learning, with sustainability and creativity at the centre. Children and adults are invited to explore, make, and learn in creative, intuitive, and imaginative ways, engaging with a thought-provoking series of experiences in an inclusive space.

Andrew Grant, director of Grant Associates and co-founder of Forest of Imagination, often reflects on how the spaces we learn in are essential in the development of our own human creativity, curiosity and confidence. Andrew proposes that every community should have its own Forest of Imagination. Andrew suggests we all need wildness in our daily lives to feed our imagination, places to escape to, to think and imagine. We inhabit a space of possibility and radical imagination. Immersed in the forest, we are sensitised, we tune in. This attunement invites us to explore radical ideas and imaginative solutions for alternative futures. Forest of Imagination is a metaphor; it creates spaces to imagine in nature. The Forest is infinite and illuminating, full of freedom and potential. Our imagination invites a response to nature and beauty; it gives us insight and guides us to dream, to be intuitive, sensitive, and open to possibility. Together, we can imagine a world that is greener and fairer, creating time and space to imagine more hopeful futures, for people and the planet to flourish.

> We need to rewild ourselves before we can rewild the planet. We need to create places in cities that inspire and feed the future creativity of our children. We need to make places where you feel part of the natural world and where you feel and see the urban landscape working as a piece - holistic, balanced, and alive.
>
> Here Forest is the home of Imagination. Imagination is everyone. (Andrew Grant, Forest of Imagination, 2024)

We have the opportunity to conceptualise how we nurture children's natural drive to learn, especially through their imagination, the arts, and in the natural environment (Hay, 2023). Care, compassion, and empathy, developed through respectful relationships, are essential conditions for creative learning, happiness, and well-being. Habits of mind are developed through learning in and about the arts—to be curious, playful, intuitive, and sensitive. Through being able to express themselves in *a hundred languages* (art, dance, movement, drama, music, writing, (Edwards et al., 1998), children's imaginations are stimulated, and they learn to be even more creative. Many of the attitudes, values, and skills that we learn in childhood stay with us for the rest of our lives.

How Far Have We Wondered?

Learning from the Living Tree Forest Mirror Maze

Forest of Imagination was invited to reimagine the Egg Theatre after it was shuttered by the pandemic. This became the first iteration of the Living Tree Forest in 2021, designed by Berlin-based artist and filmmaker Andrew Amondson and dedicated to the legacy of the late Sir Ken Robinson, patron of House of Imagination, for 20 years. A film detailing the design of the Living Tree Forest (Amondson, 2021) was shared at the international Imagine If Festival in March 2021.

Forest of Imagination also integrates the philosophy of School Without Walls started in 2010, in partnership with the Egg Theatre. This approach has invited an alternative, creative approach to learning alongside artists and creative professionals. School Without Walls is also a radical creative learning approach that challenges current orthodoxies and informs a shift in the culture of learning, placing children's agency and engagement in the arts at the heart of the process (Hay, 2021a). School Without Walls offers a creative enquiry-based approach to learning, reconceptualising the whole curriculum with the city as a campus for learning. The experience of co-designing the whole curriculum prioritises children's agency in learning. The creative methodology of the partner charity, House of Imagination, *everyone is an artist* (Hay, 2023), aims to engage everyone in their own creative expression and make this creativity visible in experimental sites for learning. This approach is embedded in the innovative creative education work of the Egg Theatre and Forest of Imagination, setting the stage for bringing a forest into the theatre.

Living Tree Mirror Maze, the second iteration of the Living Tree Forest, was a living installation at the Egg Theatre in 2022, co-created with lead artist Andrew Amondson alongside mathematician Professor Alf Coles, University of Bristol, and Marcus Rothnie, architect at Feilden Clegg Bradley Studios. The Living Tree Mirror Maze invited visitors and participants to walk through a geometric maze of mirrors into a living forest of sound, light, and sensory experiences. The sound installation by musician Cosmo Sheldrake was composed using nine years of recorded, amber-listed British birdsong, creating a beautiful and haunting soundscape for the Living Tree Forest. Designer Matthew Leece created a FUNgal network of connecting and communicating tubes to represent the concept of a mycelium network in the Egg Theatre café, inviting visitors into a deeper understanding of how the forest and humans are all connected, and to learn like a forest (Bateson, 2023).

The Living Tree Forest Mirror Maze, recipient of the Inspiring Future Generations Award 2022, was co-created as an immersive multi-sensory installation inviting reflection in the forest: art holds 'the mirror up to nature' (Shakespeare, Hamlet, Scene 3, Act 2). Together, artists and designers were engaged in a unique exploration of creative practice as a form of thinking (Manning & Massumi, 2014) *thinking through* a range of creative practices, revealing how thinking and making are intimately entangled.

In relation to ecological imagination, understanding ecologically refers to the three ecologies, the union of mind, body and environment (Guattari, 2008). Imagining ecologically is a way of paying attention to and making sense of the world that is attuned to relationships, connections, and context (Judson, 2015). Reflecting on the theories of Manning and Massumi, Robinson and Guattari, these ecologies of practice illuminate one another and help map the creative processes in Forest of Imagination, echoing the interwoven themes of creativity, the environment, and education.

Living Forest Mirror Maze invited enquiry into the natural world and supported the agency for action in response to the climate emergency. House of Imagination liaised with artists to provide a series of creative workshops for local schools in the Living Tree Mirror Maze. The installation provided a 'living classroom', an experimental site of learning. This ecology in practice highlights the emergent and entangled practices of artists, educators, and researchers.

The children related their thoughts on nature to themselves, endowing nature with qualities they valued or with human characteristics: 'Branches all connect to each other like friends'; 'If you talk to it will it answer you?'; 'Trees can grow arms'; 'Can you speak to the trees like on the first floor?' … They felt the plants, touched and stroked the mosses and ferns covering the Living Tree and examined the logs and the pool of water with their fingers and hands.

> The Living Tree Forest ... is not a heavy hitting lecture on the importance of nature, it's an experience of the joy that nature has to offer. And I think this is what art does, good art does, it teaches us about others, other people, other places, other things, other natures, and we can start to understand the degree that we all depend on each other. In the really, rough tough years that lie in front of us, and in particular that lie in front of our children, we've really got to be able to understand each other and the world we live in, and this is just a fantastically fun, beautiful, inspiring, exciting place to start. (Heffernan, 2022a, b)

Learning outside the classroom, with the city as a campus for learning, invites a new space for conversations about our collective imagination. Forest of Imagination invites us all to explore our deep human connection to nature through our imagination. Forest of Imagination encompasses educational, creative, and cultural approaches that are responsive to the natural environment through local engagement, but with a wider impact on society.

Collective Creativity

Like a forest, we are all connected (Amondson, 2020). We are connected in ways that we cannot see, yet given the right environment, it is possible to feel it. The Living Tree Forest was an experiment in environmental design, a social sculpture (Kurt & Sacks, 2013) emerging from the collective creativity of Forest of Imagination. Artists, architects, educators, researchers, environmentalists, and volunteers came together to create a forest inside a theatre. The Living Tree Forest was then animated by artist educators, who offered playful interventions to children of all ages and a range of special needs, deepening their connections not only to nature in the space but to one another. In this new '*space of liberated learning*' (Cambridge Curiosity and Imagination, 2020), synchronised synapses seed symbiotic thinking and ideas. An entanglement of empathy and inquiry creates a '*space of possibility*' (Hay, 2021b). New stories are told: a sense of belonging creates a safe space for our collective creativity to flourish.

> On entering the Living Tree Mirror Maze there was astonishment: '*Wow, I can't believe this*'. There were looks of amazement and gazing around the whole space. Reflecting on their experience of it they described how it made them feel: *amazed, calm, enchanted, magic, welcoming, warm, sleepy, summer,*

> *home, ecstatic, happy, energetic, frightened, free, damp, beautiful, good, loving, lovely, excited, thankful.* There was smiling and laughter. Living Tree Mirror Maze Case Study 2022.

The experience of the Living Tree Forest is multi-sensory; you can smell the forest before you see the forest. Your eyes scan the space to make sense of the scenery of mosses and ferns and the twisting geometry of the Living Tree. The sound of birdsong invites you to look up. We experience life in all its wonder through all our senses which are stored in our emotional memory banks. Far from the prescriptive learning often administered in the traditional classroom, these emotional connections, when accessed through open-ended multi-sensory environments like the Living Tree Forest, can create new neural pathways for learning, an embodied learning experience that can accompany children through the rest of their lives.

A growing body of research suggests that children who experience awe-inspiring moments, intrinsic to art and nature, can have long-lasting benefits, with profound impacts on their development. Awe experiences have been linked to enhanced empathy and emotional well-being, creativity and cognitive flexibility, resilience, and coping skills. But maybe most importantly, when considering ecological education, children and young people who encounter wonder and awe in natural contexts often develop a deeper connection with and appreciation for nature. This connection encourages 'environmentally responsible behaviour and a lifelong love for the outdoors. By nurturing a strong connection with nature, children are more likely to become advocates for conservation and sustainability in their adulthood' (Koehler, 2023).

Children's interpretation, debate, and thoughts on a key question posed by the artist were documented in the *Living Tree Mirror Maze film (2022):*

> Andrew Amondson:
>
> What if you could plant a tree that could save the world? And what if that tree was you?
>
> Children's responses:
>
> How could you plant a tree and that tree be you?
>
> It's a metaphor.

> You are the seed… or the idea you have is the seed of a tree and the tree grows and the branches entwine, kind of like this, and every year it gets better and better.
>
> You don't need a whole country, you don't need a whole continent, you don't need…
>
> You just need a seed
>
> You need a seed and you can change the world.

Children's voices and responses throughout the research have been key to informing each iteration of the Forest of Imagination. We have returned to the key themes of creativity, the environment, and education through an entangled approach to thinking through making, working alongside artists and educators in a new space of possibility. The focus of the Forest of Imagination 2023 explored this idea more deeply.

Assemble in the Forest

In the Summer of 2023, Forest of Imagination was invited to take over the National Trust Bath Assembly Rooms. '*Assemble in the Forest*' transformed the Grade 1 listed building into a landscape of the imagination, experienced by more than 10,000 visitors. Themed around 'Assemble', Forest of Imagination 2023 invited visitors of all ages to connect with the more-than-human world and our symbiotic relationships with it. We share over 50% of our DNA with the forest. We are nature, and nature is in each and every one of us.

Wondering Ways, co-created by Andrew Amondson and Feilden Clegg Bradley Studios, filled the Grand Ballroom with tall grasses. Visitors walking into the ballroom walked out into a large open meadow with the summer scent of 1000 potted grasses, creating paths to wander through and wonder, finding both shared and private spaces for reflection. Birdsong and music filled the air. A henge made from a fallen beech stood at one end of the Grand Ballroom. The tree, 250 years old and the same age as the Bath Assembly Rooms, was reimagined as a curving bench honouring the tree and the building's shared history. On top of a grassy hill on the other end of the room, a young sapling beech tree grew up into the chandeliers above, under its roots a small dark den of discovery where children of all ages could hide and secretly look out onto this imagined landscape.

Over the course of a month, the space was used for a number of creative interventions for schools and visiting groups. From young children sitting in a circle on the soft cork ground learning songs about the weather, to teenagers clustered on stumps talking as the setting sun ignited the tops of the grasses, to an 80-year-old person ducking, and going into the den to see what the world would look like if they were a bug. Wondering Ways became an inclusive space for collective creativity and liberated learning.

> One child who had found it hard to engage in a workshop initially, responded in a surprising way. When walking out of Wondering Ways, in conversation with the Forest team, [the child] said, 'this is the best school', and emphatically as he gazed all around him 'this is MY school. *Wondering Ways* Artist Andrew Amondson observed, 'The older generations walking into the transformed ballroom for the first time gasped with delight and clapped their hands… maybe it's because they have a fixed idea of the space and its historical context'
>
> Field notes, Assemble in the Forest Case Study 2023

Cultivating Care

What happens when you grow a living forest inside a darkened theatre or fill a grand ballroom with tall grasses? With both the Living Tree Forest and Wondering Ways, we have observed how bringing nature inside into the context of our cultural institution can help us connect more deeply with nature while at the same time having positive physical and psychological effects on people. Exposure to biophilic design has been shown to restore affect, enhance cognitive performance, reduce stress, enhance creativity, and positively impact mental health and well-being (Sharam et al., 2023).

By framing nature in the buildings that hold our stories and our histories, we foster a new relationship with it. We start to see nature as not being outside of us but part of who we are. The simple act of watering and caring for the plants, ensuring they will survive the imposed conditions of the installation, cultivates care, bringing us back to our original purpose to be good stewards of the earth. After the installation, the living plants go back to the forest or are planted as an offering of reciprocity, restoring ecologies and our commitment to them.

We have witnessed the democratising effect these living installations can have. Standing in the middle of *Wondering Ways*, one woman said, 'I never felt comfortable in places like this but now, because of this installation, I feel like I belong… I feel like it belongs to me.' To understand the relevance of wilding the Bath Assembly Rooms, it is important to understand their historical significance. It was a gathering space for Britain's most powerful and influential families; Jane Austen (nineteenth-century English novelist who lived in Bath) danced in the very same ballroom, while next door, colonial policies of extraction were being shaped over a game of cards; policies that would lead to global deforestation, ecological collapse and the climate emergency we now face. *Assemble in the Forest* was an act of reparation on behalf of the more-than-human world (jagodzinski, 2018), a great imagining of what our future might look like if we worked with nature and not against it, with the hope that the seeds of ideas born from this coming together would spread far and wide, taking root around the world.

Conclusions: Who Will Finish Our Rhymes?

Forest of Imagination is an invitation to explore everyone's creativity and imagination and to make these processes visible in the world, making a real and lasting difference to children and nature. We engage with contemporary creativity, imagine new possibilities for the way we live, and create thoughtful and imaginative spaces for everyone to enjoy. This brings a message of hope for children and young people for their rights and their future.

> At the heart of everything here is an unwavering trust in children and young people—in their creativity, their ideas and their voice … they are equal partners—not yet in their knowledge, power or capabilities but in their dignity and integrity as human beings. … Forest of Imagination is a kind of selfless art, pointing towards one of the pathways that might lead us out of the political, ethical and artistic impasse it seems to me we're caught in. (Matarasso, 2023)

This thought-piece addresses the key principles of Forest of Imagination—that it invites inquiry and elicits children's thoughts and feelings on climate change, sustainability, and positive actions for change. Documentation of the children's immersive learning experience in this 'living classroom' demonstrates the value of this experimental site of learning, as a space of possibility. However, there are some challenging questions emerging for increasingly ecological, imaginative, and aware education. Inclusive participation, ecologies of practice, and discursive processes for our team of artist educators working with an ecological approach to learning, shift our collective thinking and imagination, to enable more inquiry-focused learning in more-than-human environments. We want to rebalance climate and social justice and to enable culturally inclusive approaches to nurturing ecological imagination for regenerative learning.

We are all born with an innate sense of wonder and the capacity to imagine the impossible. What if we invited children to wild their classrooms to create a living landscape for them to care for and learn from? Creating a sense of awe and wonder helps to create empathy and acceptance of the other, both human and non-human, as well as a feeling of belonging and connectedness. We invite everyone to share experiences in the Forest of Imagination, where all ideas are valued, with no judgement or pressure; it is a space for creative play. Everyone belongs to the forest, everyone is welcome, and everyone is connected through nature and art.

> Art engages human experience, art works with human experience. Art changes human experience. (Noë, 2023, p. 162)

Together with artists and educators, we can reimagine our approaches to ecological and imaginative arts education to ensure more hopeful and creative futures alongside our young people. Imagine the possibilities. Let's set our imagination free in the wilderness; let the imagination be wild, full *of joy and love.*

Competing Interests (4) The authors have no conflicts of interest to declare that are relevant to the content of this chapter.

References

Amondson, A. (2020). *Virtual forest invitation* [Film]. Retrieved May 3, 2024, from https://vimeo.com/user10398881/

Amondson, A. (2021). *Living tree forest* [Film]. Retrieved May 3, 2024, from https://vimeo.com/user10398881/livingtree2021

Amondson, A. (2023) *Wondering ways* [Film]. Retrieved May 3, 2024, from https://vimeo.com/user10398881/wonderingways

Bateson, N. (2023), Combining, Triarchy Press, Axminster

Cambridge Curiosity and Imagination. (2020). *A call for spaces of liberated learning.* Retrieved May 3, 2024, from https://www.cambridgecandi.org.uk/resource/call-spaces-liberated-learning

Edwards, C., Gandini, L., & Forman, G. (1998). *The hundred languages of children: Advanced reflections*. Ablex Publishing.
Forest of Imagination. (2022). *Living tree mirror maze* [Film]. Retrieved May 3, 2024, from https://vimeo.com/amanifilms/forestofimagination
Forest of Imagination. (2024). Retrieved May 3, 2024, from http://www.forestofimagination.org.uk/
Guattari, F. (2008). *The three ecologies*. Continuum.
Hay, P. (2021a). *Impact case study in creative pedagogies REF21: Developing and embedding innovative creative pedagogical approaches in creative arts education*. Retrieved August 12, 2024, from https://results2021.ref.ac.uk/impact/cc2a55dd-4eb8-4ced-a38f-79849c70661c?page=1
Hay, P. (2021b). Reimagining learning: Spaces of possibility. *International Journal of Art and Design Education, 41*(1), 39–49.
Hay, P. (2023). *Children are artists: Supporting children's learning identity as artists*. Routledge.
Heffernan, M. (2022a). *Margaret Heffernan's speech at opening of Living Tree Mirror Maze*. Retrieved May 3, 2024, from https://vimeo.com/753957453/
Heffernan, M. (2022b). *Opening speech: Living tree mirror maze*. Retrieved May 3, 2024, from https://vimeo.com/753957453/4e14cc7fec?share=copy
Imagine If Festival. (2021). *Living tree*. Bath School of Design [Film]. Retrieved May 3, 2024, from https://vimeo.com/user10398881/imagineif
jagodzinski, J. (Ed.). (2018). *Interrogating the Anthropocene: Ecology, aesthetics, pedagogy, and the future in question*. Palgrave Macmillan.
Judson, G. (2015). *Engaging imagination in ecological education*. Pacific Educational Press.
Kimmerer, R. W. (2015). *Braiding sweetgrass*. Milkweed Editions.
Koehler, J. (2023). 11 benefits of experiencing awe in childhood: The benefits of unlocking awe. *Psychology Today*. Retrieved January 9, 2024, from https://www.psychologytoday.com/gb/blog/beyond-school-walls/202303/11-benefits-of-experiencing-awe-in-childhood
Kurt, H., & Sacks, S. (2013). *Make the secrets productive: Aesthetic education for sustainability*. Verlag fur Akademische Schriften.
Manning, E., & Massumi, B. (2014). *Thought in the act: Passages in the ecology of experience*. University of Minnesota Press.
Matarasso, F. (2023). *A selfless act: Trusting the child*. Retrieved May 3, 2024, from https://arestlessart.com/2023/06/25/trusting-the-child/
Noë, A. (2023). *The entanglement: How art and philosophy make us what we are*. Princeton University Press.
Paulsen, M., jagodzinski, J., & Hawke, S. (Eds.). (2022). *Pedagogy in the Anthropocene: Re-wilding education for a new Earth*. Palgrave Macmillan.
Robinson, K., & Robinson, K. (2021). *Imagine if*. Penguin.

Sharam, L. A., Mayer, K. M., & Baumann O. (2023). Design by nature: The influence of windows on cognitive performance and affect. *Journal of Environmental Psychology, 85*, 101923. Retrieved January 12, 2024, from https://www.sciencedirect.com/science/article/abs/pii/S0272494422001682 accessed 12 January 2024.

State of the Environment Report. (2021). Retrieved January 12, 2024, from https://www.statista.com/chart/27805/indigenous-communities-protect-biodiversity/

PART V

Artists and Arts Collectives: Taking Action in the Earth Crisis

At this stage ecologies in practice bring in the voices of international arts collectives and artists who are making work that helps audiences and participants learn how to make ecological changes in their lives. The artists' collectives that contribute to this section introduce their specific projects and their particular approaches to challenging the wastefulness of the Anthropocene. The projects included here are aiming for zero-waste productions, in which all materials and processes involve the imaginative recycling and repurposing of resources and energies.

The Vienna-based artists' collective WochenKlausur creates social interventions that aim to improve the quality of coexistence and reduce wastefulness. In this chapter, members of the collective present their argument for the artworld to become more accountable for its use of natural resources. The collective relates to projects that provide alternatives to exploitative systems, jet-set curation and greenwashing. It also presents projects that use the potential of art to intervene in socio-political and environmental grievances. WochenKlausur introduces initiatives that have reintegrated the capacities of local people. They have also involved students in ecological research, engaging them in rehabilitating materials wasted by theatres for imaginative upcycling.

Roberto and Juan Pablo Zamora Figueroa are film makers based in Berlin who have taken ecological film making to the extent of every possibility they can realise. The Zamora Figueroa brothers argue for film production and distribution processes to become sustainably aware of the Earth Crisis in all decisions made, from wardrobe to location and lighting, so that every element becomes waste-free, mitigating carbon emissions and caring of the Earth's ecosystems. The Zamora Figueroas invite us to accompany them in visualising a day on a sustainable film set. They also share their guide for Earth-centred film making as a transferrable code of practice for future international ecological films.

Jenn Steverson makes artworks that enable the knowledges of indigenous and historically enslaved peoples to be communicated in contemporary approaches to climate change and its causations. Steverson has responded to global warming in Texas as 'weirding' that needs to be realigned with powerful and medicinal approaches to assisting local strategies. Steverson works with archival materials and makes 'mojo bags' for climate change, that include seeds of plants that indigenous and diasporic peoples have found to be soothing and to aid survival in emergencies. Steverson argues that the waste colonialism controlled by technocrats needs to learn the lessons of history, and to learn from those it marginalises.

Authors here are offering findings made in arts practice, and with collectives who have realised projects in sustainable visual arts, eco-design, purposeful community arts and ecological technologies in film making. Each chapter contributes to a gathering of views on how artists and arts collectives can learn how to take action in the Earth Crisis, and share insights that support global regeneration.

CHAPTER 15

Art's Waste: The Wasteful Handling of Art's Potential and Resources

Martina Reuter

Now something has happened again (Haas, 1996, 2003, p. 5)

Wolf Haas, an Austrian writer, opened his early crime novels by letting the narrator say that something has happened again. The same opening would work well for the almost daily news about the consequences of the ecological system's exploitation. These consequences also often have criminal causes, but unfortunately are not fictional. Every day something happens again. Wild-fires as floods become more frequent and stronger, species are becoming extinct. CO_2 emissions are still increasing (IEA, 2024) while all kinds of waste are growing (Alves, 2023).

When writing this chapter, February 2024 was announced the warmest February since weather recording, 'making it the ninth consecutive month with a global temperature record' (Copernicus, 2024). Austria also

M. Reuter (✉)
WochenKlausur, Vienna, Austria
e-mail: wk@wochenklausur.at

M. Matthews (ed.), *Ecologies in Practice and Learning: Arts Interventions in the Earth Crisis*, Palgrave Studies in Educational Futures, https://doi.org/10.1007/978-3-031-82338-1_15

recorded an ‘Overshoot Day’ on April 7, 2024 (Earth Overshoot Day, 2024). This means that if everyone on Earth lived like Austria’s obviously wasteful population, all renewable resources for the current year, such as energy and food, would be used up worldwide by April 7, 2024. Luxembourg, by the way, had already had its Overshoot Day in February 2024.

How can such dynamics ever be brought into balance? Early solutions to the major problems are not in sight despite decades of urgent warnings from the scientific community, despite startling news and documentaries, despite actions by climate organisations and activists and international climate summits: little is changing. And if you think about the dominant topic of all the art exhibitions, art fairs and art magazines, which has been hyped for some time now, namely the environmental and climate crisis, you could even think despite the art world’s massively exhibited concerns, is the art world itself an exceptional sector when it comes to wasting resources? Is waste caused by the art itself not significant? Does the art world care about re-use or recycle its own waste?

Certainly, art will hardly become the saviour of the Earth, art will not by itself make the mountains of waste disappear and will not be able to stop the waste of resources such as energy. But if artists become more aware of their potential to actually change things in the real world, their contribution to solving problems could be a little more effective.

The following chapter aims to take a brief look at the art sector’s possible part in the ecological dilemma, but also to highlight a few of the many exemplary projects by ecologically and socially committed artists who have played their part in improving real conditions, taken on ecological responsibility or recognised ecologically committed art as an educational tool. The aim is not to retell a historical-chronological narrative, but to show some of the diverse thematic approaches of art that describes itself as ecologically, socially or politically dedicated.

Art’s Wasteful Use of Its Own Potential

To make a decisive difference to a sustainable turnaround, it will probably not be enough to use ecologically compatible paints or canvases. The request is that the art world’s theming of climate, ecology and the environment should be followed by more far-reaching undertaking. The credibility of ecological and social criticism in artistic representations and performances must be put up for discussion.

With his article, 'What the warming world needs now is art, sweet art,' the American journalist William McKibben (2005), who won the Right Livelihood Award (2014), complained that the topic of global warming did not play any role in culture and art:

> Here's the paradox: If the scientists are right, we're living through the biggest thing that's happened since human civilization emerged. One species, ours, has by itself in the course of a couple of generations managed to powerfully raise the temperature of an entire planet, to knock its most basic systems out of kilter. But oddly, though we know about it, we don't *know* about it. It hasn't registered in our gut; it isn't part of our culture. Where are the books? The poems? The plays? The goddamn operas? (McKibben, 2005)

In contrast to the enormous engagement of artists and their political impact during the AIDS pandemic in the 1980s and 1990s, a commitment to the environmental crisis was missing, much to McKibben's regret.

Many people were threatened and affected by the pandemic at the time, including many artists who protested against the irresponsible HIV policy, and this is certainly a reason for art's intensive engagement (Lowery, 2022). But in 2005, when McKibben published his article, many people seemingly still believed climate change would not affect them. Nor was the public really aware of art that had addressed ecological problems long before. He nevertheless came to the conclusion that it was better to let the facts speak for themselves: 'If we can't turn to creative artists, then to documentarians' (McKibben, 2005).

Art's Wasteful Use of Resources

Has William McKibben's wish for a 'goddamn opera' critically dedicated to climate catastrophe finally come true, 14 years after his appeal to the artists? In 2019, the Venice Biennale was all about climate change and the Lithuanian artists Vaiva Grainytė, Lina Lapelytė and Rugilė Barzdžiukaitė received the Golden Lion for their eco-critical opera performance Sun & Sea (Marina). The opera performance, which has travelled all over the world, shows a stage scene with more than 20 performers basking in the "sun," singing, eating, walking, playing on cellphones at an "indoor beach," set up from tons of delivered sand, using energy for light, audio, and in some venues heating equipment. Visitors can watch the opera from gallerias or balconies above. As art critic Agnė Sadauskaitė (2018) said,

> Subtly but ironically, the three artists chose to segment narratives about climate change, the exploitation of natural resources and ecology, in a piece in which the cult of human beauty and the mutability of the body correlates with the pain and fragility of the Earth. (Sadauskaitė, 2018)

The Earth's fragility will certainly be subtly reflected in the opera by the artists, but the image that they present of themselves while critically holding up a mirror to society could be subtitled with the question: must we practice what we preach? Perhaps the answer has already been found by Joshua Barone of The New York Times, when the opera was performed at the Brooklyn Academy of Music in 2021:

> What they didn't want was to give the impression that they were climate activists. 'It would be unfair to say that' Grainytė said. 'If we were activists, we wouldn't create this work that is travelling the world' (The production, like many in the performing arts, isn't the most Eco-friendly: For the BAM presentation, all that sand was transported by truck from Volleyball USA in New Jersey to Brooklyn). (Barone, 2021)

Barone states in his article, roughly speaking, that the creators of Sun & Sea (Marina) do not intentionally evade responsibility: politically engaged artists are aware of grappling with difficult issues, they just want their opera to 'activate,' as Lapelytė expressed it.

The audience to which art projects like this opera performance are aimed might actually be assumed to be already well informed about the ecological crisis and therefore already activated to lead a more ecologically conscious life. This is by no means to question the effectiveness of ecologically critical art, but if its implementation does not do justice to its own criticism of waste and exploitation of resources, the question arises as to whether such works can really be considered responsible.

However, according to the numerous benevolent, sometimes enthusiastic reviews from art critics, the opera performance Sun & Sea (Marina) seems to be fascinating, astonishing and beautifully staged and is adored by the audience. Who knows, maybe it will at least stop some visitors from wasting resources by flying to another beach paradise.

Yet there are still facts of the busy art sector that make it difficult to take their ecological concerns completely seriously, despite ambitious undertakings such as those of the Gallery Climate Coalition (2024), an "international membership organization providing environmental sustainability guidelines for the art sector," and announced efforts that have been increasingly circulating on various websites, such as from Art Basel (2024) or from Frieze (2024) to carry much more about sustainability in the future. The ecological footprint of the art world cannot yet be negated. Artworks are regularly transported all over the world, as well as stage equipment, countless curators, critics, collectors, gallery owners, museum directors, artists, orchestras, actors, filmcrews, managers, and visitors constantly fly to international exhibitions, all kinds of festivals, art fairs and auctions. Art critic Hanno Rauterberg stated in Greenwashing, Die Kunst der Scheinheiligkeit (Greenwashing, the Art of Hypocrisy, 2020).

> It is taken for granted that curators jet halfway around the world for a short studio visit, that artworks are constantly being sent by express airmail and that the airports at the fairs in Miami or Basel are practically clogged because so many collectors arrive by private jets.

If an ecological critique of the art world is meant seriously, then it must surely be in line with the ecological behaviour of its stakeholders. There are examples that demonstrate that this is possible.

Art's Useful Use of Waste

When it comes to ecologically responsible art movements, Helen Mayer Harrison and Newton Harrison were pioneering role models for later artists also working in ecologically engaged practice. This is especially so because they were forerunners in investing their artistic creativity in research and working on realisable projects to improve the environment. Their first project began in 1970: Earth Making was not only the title but the description of the action (The Harrison Studio, 1970). Newton Harrison produced 'earth', by using industrial by-products, kitchen waste, sawdust and sand. The aim was not to use it in a sculpture or as part of an installation, but simply with the intention of producing earth. Although aspects of a performance can be recognised in the Harrisons' process, as documented in photography, the project continued for several years and this examination of earth made the theme of endangered and polluted topsoil its central theme.

After further research projects, the Harrisons, as they were often called, were invited in 1989 by the director of the Berlin Botanical Garden in order to help with the establishment of the nature reserve in a floodplain area of former Yugoslavia. On site the artists saw that the ecological system was extremely endangered by sewage contaminated from industrial waste. For their project Breathing Space for the River Sava, they put their examination of the history and condition of the region, and their reflections on their project into the form of a poem. But the central part of the project was the development of a feasible plan for a nature corridor kept from any waste-endangering along the Siva River (The Harrison Studio, 1989).

The Harrisons dedicated their entire artistic careers of more than 40 years to comprehensive ecological research projects and their implementation plans, for which they collaborated with countless professionals from diverse fields such as eco-science and sociology, including local people to explore the socio-ecological impact of life-supporting systems.

Reusing wasted or discarded material with the intention of reducing waste and at the same time creating objects of higher value out of the waste, called upcycling, is another idea being pushed by some artists, designers and architects. The term upcycling was used first in 1994 by the engineer Reiner Pilz, who was interviewed for Salvo—an initiative and platform which was launched in 1991 by Thornton Kay (1994) in the UK, originally as a monthly magazine for the reuse of architectural salvage.

> We talked about the impending EU Demolition Waste Streams directive. 'Recycling' he said, 'I call it down-cycling. They smash bricks they smash everything. What we need is up-cycling where old products are given more value, not less. (Kay, 1994)

Starting from this understanding of reusing wasted material, WochenKlausur, the artists' collective to which the authors of this article belong, were however inspired by the ecological art movement, and after having already implemented some socially engaged projects, the collective initiated its first project with an ecological aspect in 1998. WochenKlausur set up a model for the local and regional reuse of waste in the sense of upcycling in Linz, Austria. After meetings with experts from the fields of ecology and design, and local companies, some ideas were put into action. For example, an Austrian brewery commissioned the designer Susanne

Lipschitz who developed a process for reusing defective beer cans for producing floor tiles. Some of these were used in the brewery's taproom as an illustrative sample (WochenKlausur, 1998).

Art's Potential: Too Useful to Be Wasted

It is clear that the ecological and social relevance of art can be interpreted in many ways. Some artists have for several decades now decided to use art's potential to intervene in a corrective way in the ecological or social circumstances of life.

WochenKlausur has also seen art as a tool to change things for the better, since their first socially engaged art projects (Hawley, 2015, pp. 83–109). The artist's collective planned to establish a mobile healthcare provision for homeless people in Vienna in 1993. Among other tasks, it was important to get the city health councillor's agreement to finance the salaries of the doctors for the planned mobile medical practice. After the first negotiations without success, the artists persuaded a journalist to conduct a fictitious interview with the councillor about the project. Immediately after the "interview," the city council agreed to provide the requested funding (WochenKlausur, 1993).

For the necessary financial support of the Zurich project in 1994, aimed at establishing a shelter for drug-using and homeless female sex workers, the key representatives of all political parties, of the judiciary, and of the police were asked to participate in several small group discussions with representatives of drug counselling and the affected individuals themselves. For two weeks, small groups of four to five participants were invited to small boat trips on Lake Zurich—also to prevent them from leaving the conversation prematurely, but primarily to discuss drug policy and the life experiences of those affected away from the public eye, and thus without the pressure of public opinion. The result was the financial commitment to the project (Kester, 2004).

However, after those projects were implemented and were then frequently the subject of panel discussions among artists, art critics and the art public, WochenKlausur was often accused of not doing art but social work. The collective in fact sees itself as offering possible opportunities to learn about and correct socio-political deficits. Not wasting this possibility of art, namely to shape real living conditions, is a premise for the group when developing projects.

Some shortcomings can often be remedied more easily and quickly by people who are not involved in the bureaucratic systems associated with the issue in question. They can bypass official and hierarchical structures more easily, shorten official channels, and negotiate with authorities more impartially with an outside perspective, in comparison to those who are integrated into the official system as clients, as staff or as experts. If artists use the principle of artistic freedom for artistic strategies, the institutional capital of commissioning art and cultural institutions or companies and, last but not least, their artistic creative potential, a lot can be achieved. It would have been almost impossible for a social worker to set up a mobile medical practice and secure its long-term financing within 11 weeks. For WochenKlausur as an artists' collective it was feasible.

Artists are usually not social workers, but with their specific possibilities they can help to enhance certain aspects in the social field. Admittedly, artists are usually not ecologists either, but they can contribute to improving ecological conditions with their alternative perspectives on the matter, and with methods and possibilities that are perhaps unfamiliar to science.

When artists decide to dedicate their work to ecologically endangered and socially vulnerable areas, this must of course be accompanied by responsibility. Learning from people's knowledge and involving them is a basic condition, as is offering the specific capital of art as an instrument to achieve the jointly defined goals for improving the respective situation. If ecologically or socially intervening artists are not aware of all the possible consequences of their actions, the results of their projects may even worsen the conditions they promised to improve. "Sorry, the project failed, but as an artist I was able to learn from it," should not be the intended benefit of such projects.

The Argentinian collective Ala Plástica impressively showed what it is to understand how environmental art projects can be managed responsibly. Their initiative the AA Project from 2000 in the Paraná Delta region is cited here as only one example of their artistic approach (AA Project, 2000). After studying the bioregional system in the area, the group looked at the regional living conditions and learned about local traditions and environmental challenges and planned to design a map together with the locals. The aim was to use such a map to restore, in collaboration with local experts and the comprehensive local knowledge of the region, what was wasting away: the topography, habitat and agricultural traditions. Based on the information and records collected, workshops to learn about traditional crafts and organic farming were incorporated into training programmes for local communities. Another important effect of the artists' efforts was that the project enabled the joint construction of river embankments to reduce flood damage. Grant Kester refers to this project in his book, The One and The Many (2011).

Finally, How Not to Waste the Educational Value of Ecologically Dedicated Art

Now that it is a matter of at least maintaining the status quo in order to preserve this Earth as a planet that is reasonably worth living on, art, which claims to be ecologically committed, must itself behave responsibly. It must not only point out the damage but also help to limit it. The fact that many young people often see little hope for an untroubled future, as particularly evident in the name chosen by young climate activist groups in several European countries—namely Last Generation/ Letzte Generation (2024), should not prevent them from continuing to engage constructively in projects promoting sustainability and social responsibility.

An outstanding and renowned example of an ecological education project was initiated in 1977 by Israeli artist Avital Geva. The Ecological Greenhouse at Kibbutz Ein Shemer. This project originally aimed to bring together families from conflicted Jewish and Arab communities in the area, to work together on the same site. Over the years, the project has developed into an educational research centre, with a wide programme of ecological science, agriculture and technology.

> It is a place where people from the entire spectrum of Israeli society meet in order to invent, learn, research, create ideas and find solutions for a greener future through innovation, sustainability, agriculture, science, technology and culture. Here, start-up entrepreneurs meet with high school students, artists and scientists, Jews with Arabs, secular with religious, new immigrants with locals, young with old, under one roof to work on issues in climate, food, water, waste, energy, consumption, and culture. (The Ecological Greenhouse, 2024)

The Ecological Greenhouse Team illustrates the programme. This astonishing exemplary project accomplishes several ambitious tasks and is an excellent model to emulate. Even during the current difficult times since the Hamas terror attack on October 7, 2023, and the following

Israel-Hamas war in 2024 the Growing together programme is not only continuing but even expanding (2024).

In order to appeal to ecological and social awareness in arts learning programmes at universities, WochenKlausur set up a project for the exhibition Beyond Green: Toward a Sustainable Art, curated by Stephanie Smith for the Smart Museum in Chicago in 2005. While researching information that could be inspiring for this project, it turned out that the Smart Museum, as well as the temporary office of WochenKlausur at the Art Department of the University of Chicago, was situated adjacent to the neighbourhood of Woodlawn. This was one of the most socio-economically challenged districts in Chicago in 2005. It was self-evident to include this fact in the project plan—together with the ecological reference, identified in the title of the exhibition. The group knew from experience that cultural institutions like theatres, museums and galleries usually discard stage sets and materials used for temporary exhibitions—waste that can still be of use for other purposes.

These were the factors that WochenKlausur wanted to bring together: firstly the exhibition's purpose to spotlight art and design movements which were addressing environmental challenges; secondly the waste caused by art and cultural institutions themselves; thirdly the social aspects concerning the neighbourhood of the university. In addition, the group's temporary office was set up at the university's art department and, therefore, WochenKlausur met art and design students on a daily basis.

Based on these conditions, a network was implemented among the social facilities in Woodlawn that included homeless shelters, soup kitchens and more. All of these facilities would send their needs for furniture or furnishings to the art and design students at the university, who would then design and build those requested items with the materials they received from cultural institutions that were willing to regularly supply students with their otherwise discarded materials.

The final task of the project was to establish a non-profit organisation to take over the long-term operation and coordination of the network. This new organisation, called Material Exchange, was composed of staff from the Smart Museum, the University of Chicago Art Department, the Illinois Institute of Technology, and Harrington College of Design, as well as representatives from the participating social service agencies. The network was in operation until 2010, when the managing board decided to close it. For a couple of years at least, the networking system of Material Exchange showed one of many possible ways to deal with waste caused by art and cultural institutions (WochenKlausur, 2005) (Fig. 15.1).

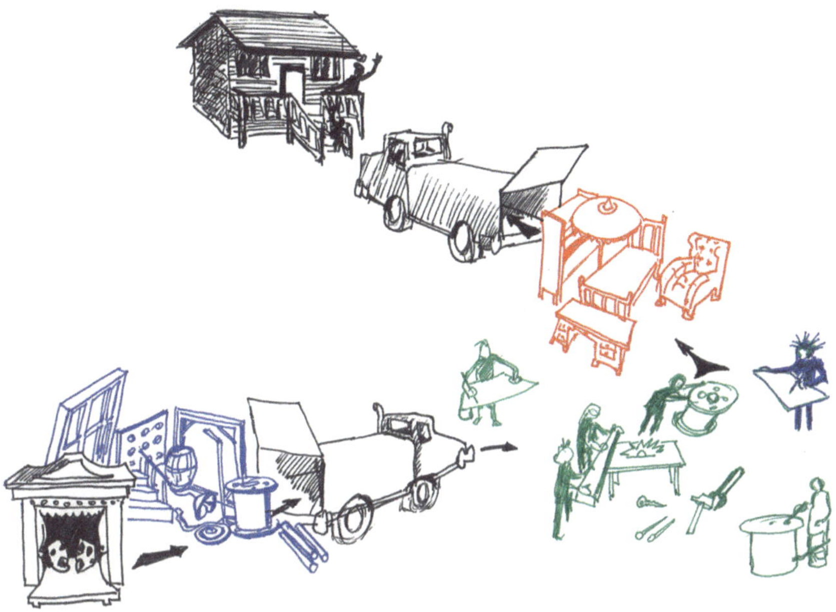

Fig. 15.1 WochenKlausur Chicago Upcycling Scheme project

Conclusion

The potential of art to contribute to improvements of ecological and social deficits should neither be misunderstood, underestimated, nor wasted. The few examples mentioned in this chapter have hopefully justified this view on art.

Art interventions like those made by the Harrisons, Ala Plastica or WochenKlausur aim to effectively change local conditions in need of improvements in a long-term way, regardless of whether they are of interest to an audience or not, whether they are successful in the art market's sense or not—but with a high level of responsibility towards all those involved in their projects, especially the locals affected. Artistic freedom enables unusual methods, for example to achieve the goals and thus the main purpose of interventions made, by means of creative negotiation strategies.

Art, on the other hand, such as the opera project Sun & Sea (Marina) by Vaiva Grainytė, Lina Lapelytė and Rugilė Barzdžiukaitė, is aimed at the attention of the art world, at its audience, which may be emotionally activated by such performances to take the environmental crisis seriously. Reminding the art sector of climate change through aesthetically expressed critique, whether in the form of images, performances, sculptures or installations, can certainly be effective, but it would be desirable if, as mentioned above, the artists also do justice to their own critique with their work.

Of course, ecologically committed artists also travel, but if they not only call themselves such but also feel committed to the environment, they will do so responsibly. And of course, most artists want to present their work, their often elaborate productions, not only in their neighbourhood; but it is time to find creative solutions that allow artists, and all interested, to participate in the art sector at least with less CO_2-intensive processes.

Finally, however, it would be worth noting that in addition to all the seriously engaged art that needs to be recognised, the fact is that even the most dedicated regional eco-project implemented by locals is never celebrated in the spotlight as much as an arts festival, and even the most committed social worker never receives the same media and public attention as a popular artist. Additionally, even the most dedicated school working to reduce social injustice for its students never receives the same appreciation as a celebrated theatre. This is by no means a call to pit art and local environmental initiatives or social work against each other, but rather to think about using a little more of the potential and capital of art for ecological and social issues—beyond well-meaning charity and donations.

Competing Interests (4) The authors have no conflicts of interest to declare that are relevant to the content of this chapter.

References

AA Project. (2000). Arte útil. Archive / AA Project. Memory database and participatory map of Delta region. Nr. 370 / 2000 - ongoing / Argentina. Retrieved May 29, 2024, from https://www.arte-util.org/projects/aa-project/

Art Basel. (2024). *Sustainability. Discover Art Basel's commitments in order to improve its environmental impact.* Retrieved May 29, 2024, from https://www.artbasel.com/sustainability?lang=de

Alves, B. (2023, December 18). *Global waste generation - statistics & facts.* Retrieved May 29, 2024, from https://www.statista.com/topics/4983/waste-generation-worldwide/

Barone, J. (2021, September 14). A climate opera arrives in New York, with 21 tons of sand. *The New York Times Online*. Retrieved May 29, 2024, from https://www.nytimes.com/2021/09/14/arts/music/sun-and-sea-bam.html

Copernicus EU. (2024, March 8). *February 2024 is the warmest February on record*. Retrieved May 29, 2024, from https://www.copernicus.eu/en/media/image-day-gallery/february-2024-warmest-february-record

Earth Overshoot Day. (2024). *Countries overshoot days 2024*. Retrieved May 29, 2024, from https://overshoot.footprintnetwork.org/newsroom/country-overshoot-days/Frieze

Frieze. (2024). *Frieze's environmental responsibility statement*. Retrieved May 29, 2024, from https://www.frieze.com/page/friezes-environmental-responsibility-statement

Gallery Climate Coalition. (2024). *Creating an environmentally responsible art world*. Retrieved April 15, 2024, from https://galleryclimatecoalition.org/

Haas, W. (1996). *Auferstehung der toten*. Rowohlt Taschenbuch.

Haas, W. (2003). *Das ewige leben*. Hofmann und Campe.

Hawley, E. S. (2015). Art, activism and democracy: WochenKlausur. *Peace & Change, 40*(1), 83–109.

IEA (International Energy Agency). (February 2024). *CO_2 emissions in 2023. A new record high, but is there light at the end of the tunnel?* IEA Publications. Retrieved May 29, 2024, from https://iea.blob.core.windows.net/assets/33e2badc-b839-4c18-84ce-f6387b3c008f/CO2Emissionsin2023.pdf

Kay, T. (1994). Reiner Pilz: Salvo in Germany. Salvo - a monthly look at architectural antiques, reclaimed building materials & allied topics. No 23 + salvo news 99. October 1994 (p. 11). Retrieved April 15, 2024, from www.salvoweb.com/files/sn99sm24y94tk181119.pdf

Kester, G. A. (2004). *Conversation pieces: Community and communication in modern art*. University of California Press.

Kester, G. A. (2011). *The one and the many: Contemporary collaborative art in a global context*. Duke University Press.

Lowery, J. (2022). *It was vulgar and it was beautiful: How AIDS activists used art to fight a pandemic*. Bold Type Books.

Last Generation / Letzte Generation. (2024). Retrieved May 29, 2024, from https://letztegeneration.org/en/

McKibben, W. (2005). What the warming world needs now is art, sweet art. *Grist. Climate.Justice.Solutions*. Retrieved April 15, 2024, from https://grist.org/article/mckibben-imagine/

Rauterberg, H. (2020). Greenwashing: Die kunst der scheinheiligkeit. In N. Sievers, U. Blumenreich, S. Dengel, & C. Wingert (Eds.), *Jahrbuch für Kulturpolitik 2019/20* (pp. 153–158). https://doi.org/10.14361/9783839444917-020

Right Livelihood. (2014). Awarded 2014. Bill McKibben, 350.org. Retrieved May 29, 2024, from https://rightlivelihood.org/the-change-makers/find-a-laureate/bill-mckibben-350-org/

Sadauskaitė, A. (2018, October 18) Signs of the apocalypse on the beach. An interview with artists Vaiva Grainytė, Lina Lapelytė and Rugilė Barzdžiukaitė. *Echo Gone Wrong*. Retrieved May 29, 2024, from https://echogonewrong.com/signs-apocalypse-beach-interview-artists-vaiva-grainyte-lina-lapelyte-rugile-barzdziukaite/

The Ecological Greenhouse. (2024) *Learning in the real world*. Retrieved May 29, 2024, from https://greenhouse.org.il/en/who-we-are/

The Harrison Studio. (1970). *Making Earth 1970*. Retrieved May 29, 2024, from https://www.theharrisonstudio.net/making-earth-1970

The Harrison Studio. (1989). *Sava river 1989*. Retrieved May 29, 2024, from https://www.theharrisonstudio.net/sava-river-1989

WochenKlausur. (1993). *Medical care for homeless people*. Retrieved May 29, 2024, from https://wochenklausur.at/projekt.php?lang=en&id=3

WochenKlausur. (1998). *Upcycling cooperative*. Retrieved May 29, 2024, from https://wochenklausur.at/projekt.php?lang=en&id=11).

WochenKlausur. (2005). *Furnishing social institutions*. Retrieved May 29, 2024, from https://wochenklausur.at/projekt.php?lang=en&id=21

CHAPTER 16

A Seed Towards Regenerative Filmmaking

Juan Pablo Zamora Figueroa
and Roberto Andrés Zamora Figueroa

Introduction

Films have been spreading environmental awareness since their creation. While some have done this by addressing the multiple climate injustices around the world, such as Chasing Coral (Orlowski, 2017) on the disappearance of coral reefs, others have imagined dystopia in far distant futures where nature has been destroyed by humans, such as WALL-E (Stanton, 2008), a children's movie about the last garbage cleaning robot left on a desolated Earth. Addressing real-world crises within a script has always helped writers to anchor movies in reality, making them more relatable. We have had documentaries about animal and soil conservation, the meat industry, dramas about colonisation, nuclear catastrophe, and multiple natural disasters. Yet every single one of these movies has been powered by fossil fuels. This chapter recognises the filmmaking industry as a significant contributor to greenhouse gas emissions: from multiple flights to shoot one nature documentary, to big studio productions composed of

J. P. Zamora Figueroa (✉) • R. A. Zamora Figueroa
Fën, Berlin, Germany
e-mail: contact@fen.earth; contact@fen.earth

M. Matthews (ed.), *Ecologies in Practice and Learning: Arts Interventions in the Earth Crisis*, Palgrave Studies in Educational Futures, https://doi.org/10.1007/978-3-031-82338-1_16

thousands of lights, one-time sets, and motorhomes for celebrities. We have come to question how some of the most ambitious movies in our culture have turned out to be as polluting as, in some cases, the industries or economic systems they are questioning (Bailey, 2010). Douglas Kellner, an American critical theoretical author, mentions in an article how some Hollywood films are showing the dangers of the environment due to out-of-control corporations in need of regulations—he argues that these films send a powerful message to those polluting companies by saying that if you make a mistake, someday, a film can be made about your mistake (Polakovic, 2001).

Exploring how little Hollywood has done so far to mitigate their carbon emissions we became prompted to challenge the status quo of filmmaking, through the creation of Fën, an ecological film collective seeking to revolutionise films by incorporating various sustainable practices.

Shifting our focus to ecological filmmaking, we share in this chapter, guidelines we have written for each film department, forming the first sustainable production manual (Zamora et al., 2013). We explore concepts such as waste management overseen by the production, renewable energy usage, cinematography advocacy for energy-efficient lighting, eco-conscious decisions within the art and catering departments, among others.

After the turbulent events of 2019 and the Covid-19 pandemic, we recognised that a unique period of introspection was prompted among us, leading to the emergence of 'eco-affectivity', a holistic approach blending ecological and emotional aspects for those involved in filmmaking. This philosophy culminated in the production of Mar & Cielo (Zamora, 2021), Latin America's first carbon-neutral short film (Goldaracena, 2023). Resulting in a minimal carbon footprint through circular ecologies, zero waste code on set, vegan local catering, and carbon offsetting, Mar & Cielo represents Fën's shift from sustainable guidelines to regenerative filmmaking.

In this chapter we recognise the profound impact filmmaking can have on environmental awareness and societal values. Embracing regenerative filmmaking, we established 11 key agreements to guide our practice, from awareness and sustainability to responsible co-creation and eco-affectivity. These agreements drive us to produce films that not only minimise carbon footprints but contribute to the healing of ecosystems and communities. As we embark on this journey, we invite you to envision a day on a regenerative film set, where every action echoes a commitment to a sustainable and harmonious future, encapsulating a holistic approach to filmmaking.

Through sharing our experiences as ecological filmmakers and our goal as ecological people, we look forward to envisioning a world where all audiovisual content is made as a regenerative practice by bringing relief to the atmosphere and also becoming part of its healing. In this vision of regenerative filmmaking, every idea is a seed, and by the final result, it is a tree healing the earth.

Going Beyond Carbon Offsetting Schemes

A 2006 study from the University of California estimated that the film industry produces 15 million tonnes of CO_2 a year, making it in places like Hollywood, the second-biggest polluter right after oil refining, more so than aerospace, clothing, and the hotel industries (Hoad, 2020). To anyone in the film sector this is no wonder. From the minute you start downloading an idea into reality there is a significant use of resources. From development, with all its consumption of paper, to the pre-production with its construction of sets, all the way to filming, which involves, among many factors, excessive use of electricity, mobilisation for cast, crew, equipment and production necessities, catering filled with bottled water, special effects, with their explosions and their invasive nature on locations, and overall loads of waste.

In our own journey to become filmmakers these findings made us, Roberto and Juan Pablo, brothers and co-authors of this chapter, question our desire to be part of the film industry. From this realisation, back in 2013, we founded Fën (seed in Mapudungun, language of the Native Chilean Mapuches), an ecological film collective, in which we agreed that to make an ecological production was beyond telling a story with environmental awareness throughout the script. Just a few months earlier Roberto, Javiera Cifuentes, and Catalina Jaque's thesis, "Film, An Ecological Activist" (Zamora et al., 2013), had formed the basis for Fën's ecological filmmaking practices. This thesis explored three key areas: the presence of ecology in cinema's history, prior attempts at sustainable film production, and a proposal of guidelines for ecological filmmaking. We wanted to make movies without making a mess.

Discoveries about carbon offsetting schemes from this research would end up informing Fën's distinct approach. Syriana (Gaghan, 2005), a film that reveals the corruption and political dishonesty involved in maintaining our addiction to fossil fuels, exemplified carbon neutrality by offsetting its emissions through renewable energy investments (Warner Brothers,

2006). Despite its significance, we noticed from the Syriana case, that although the film was a certified carbon neutral production, no one in their cast or crew continued to make carbon-neutral films, as if there had been no relevant learning experience from making the movie. It was this fact, mainly, that inspired us to start thinking about sustainable filmmaking as something beyond the money-making industry mentality of Hollywood. We decided to approach this concept from a place of radical action and creation.

As a bridge between our research into ecological activism and film, and our practical laboratory through Fën, we created our first manual for sustainable production in practices, which we applied through the making of Traffic Light (Zamora, 2014), a 30 second film made solely with three bikes, our family and friends, a second hand camera, and a solar energy grid bought after a month of working as waiters.

Sustainable Production in Practice

In 2014, through our manual we developed comprehensive guidelines for sustainable production, aimed at minimising environmental impact and fostering ecological consciousness. These guides emphasised the dual purpose of ecological filmmaking: to both produce environmentally conscious films and to inspire long-term commitment to ecosystem restoration. The guidelines were as follows:

The production department is the first line to plan and measure the impact of a production.

By calculating carbon emissions and integrating sustainable practices from the outset, such as utilising renewable energy sources and implementing waste management strategies, the department set the stage for eco-friendly filmmaking (Anderson, 2023). In pre-production, efforts were made to reduce paper usage by utilising digital communication and sharing resources electronically. When printing was necessary, eco-friendly practices such as using recycled paper and energy-efficient fonts were employed, like Century Gothic as it uses 30% less ink than the classic Arial, and remains easily readable (Carmichael, 2023).

The cinematography department embraces energy-efficient technologies, such as LED lighting, and advocates for natural lighting whenever possible to reduce energy consumption (Winstead, 2023). Similarly, transitioning from analogue to digital camera systems will minimise resource-intensive processes and enhanced sustainability (Ponti, 2022). Unlike analogue film, digital formats significantly reduce resource consumption by eliminating the need for water, paper, and toxic chemicals associated with processing. Moreover, digital media avoids the use of gelatine of animal origin, making it a more vegetarian-friendly option.

The art department adopts the principles of ecological design, prioritising sustainable sourcing, reusing, and recycling in set construction and costume design. By incorporating the six Rs of ecology (Rethink, Refuse, Reduce, Reuse, Repair, Recycle), they ensure that every element contributes to a more sustainable production (Antico et al., 2017). The costume department prioritises sustainable practices by avoiding fast fashion and opting for thrift stores to dress characters, thereby reducing the accumulation of discarded clothing in landfills, such as the significant volume seen in Chile's Atacama Desert (Duong, 2021). Additionally, the makeup department has to ensure that cosmetics are free from polluting chemicals and are not tested on animals, aligning with ethical and eco-conscious principles.

Efforts extend to visual effects, sound design, and post-production, where digital technologies and renewable energy sources are utilised to minimise environmental impact.

Moreover, the catering department prioritised eco-friendly practices, opting for reusable utensils and vegan catering to reduce waste and lower carbon emissions (Scarborough et al., 2023).

Transportation is one of the most important and difficult areas to reduce our emissions as film productions often rely on fossil fuel for transporting equipment and crew. To offset carbon emissions, the transport department aims for climate positivity, urging producers to opt for electric or biodiesel vehicles when possible, yet challenges persist in sustainable mobility. The production of electric vehicles relies on minerals linked to human rights abuses and environmental degradation (Lebreton, 2023). Alternative options like iron-based batteries are emerging but require a shift in perspective. Recommendations for this department include providing bikes for crew and incentivising sustainable commuting methods. The management of residual waste is approached systematically, with emphasis on recycling and innovative solutions like eco-bricks for unrecyclable components (Antico et al., 2017).

Through these concerted efforts across departments, our collective seeks to inspire a paradigm shift in the filmmaking industry towards sustainability. And while some of these endeavours have been adopted by some filmmakers thanks to the incorporation of Green Consultants

(Christensen, 2023), their activist nature becomes limited by the size of the production. As we continue to navigate the complexities of ecological filmmaking, we sustain the importance of fostering awareness in our crew and cast, contributing to a more sustainable future.

Emotional Well-Being as a Wheel of Sustainable Filmmaking

The commitment to sustainable filmmaking, informed by principles addressing environmental impact, societal welfare, and economic viability, reflects an evolving comprehension of the intricate relationship between humanity and the natural world (Becker & Manstetten, 2004). Historically, human activities have often exhibited a desire to dominate nature, a mindset that has contributed to contemporary environmental challenges (Loreau, 2023). The societal repercussions of this detachment became evident in the upheavals of 2019, marked by global protests and social unrest (Scheidel et al., 2020), showcasing the interconnectedness of environmental and social issues.

While looking for ways to improve on the ecology of our film practice, we came up with the idea of 'eco-affectivity', thinking of an approach that would combine ecological considerations with affective emotional aspects. By accepting that we are part of nature, we treat nature how we would take care of ourselves and each other, with kindness, love, and affection. In the context of sustainable filmmaking, it involves incorporating emotional and affective elements into our environmental practices, going beyond the mere conservation of resources to consider human emotions and social dimensions within the set.

Acknowledging the need for a more comprehensive approach for 'eco-affectivity', we found support for it through the concept of 'emotional ecology' (Soler et al., 2003), defined as the art of transforming and directing our energy in ways that have an impact that can improve our personal growth, the quality of our relationships, and promote better care of our world. It invites us to become aware that we are part of a human and natural ecosystem, within which we are all important pieces and influence the whole. This marked a conscious effort to align filmmaking practices with emotional well-being (Rowold, 2011).

Recognising the need for a comprehensive approach, we introduced the role of an eco-affectivity manager to our production sets. This role focuses on nurturing a positive team environment, managing eco-anxiety, and fostering emotional well-being (Bilican, 2016). In 2019 another seed was beginning to sprout in our garden, a long-forgotten script about a place that we had carried in our hearts since childhood (Fig. 16.1).

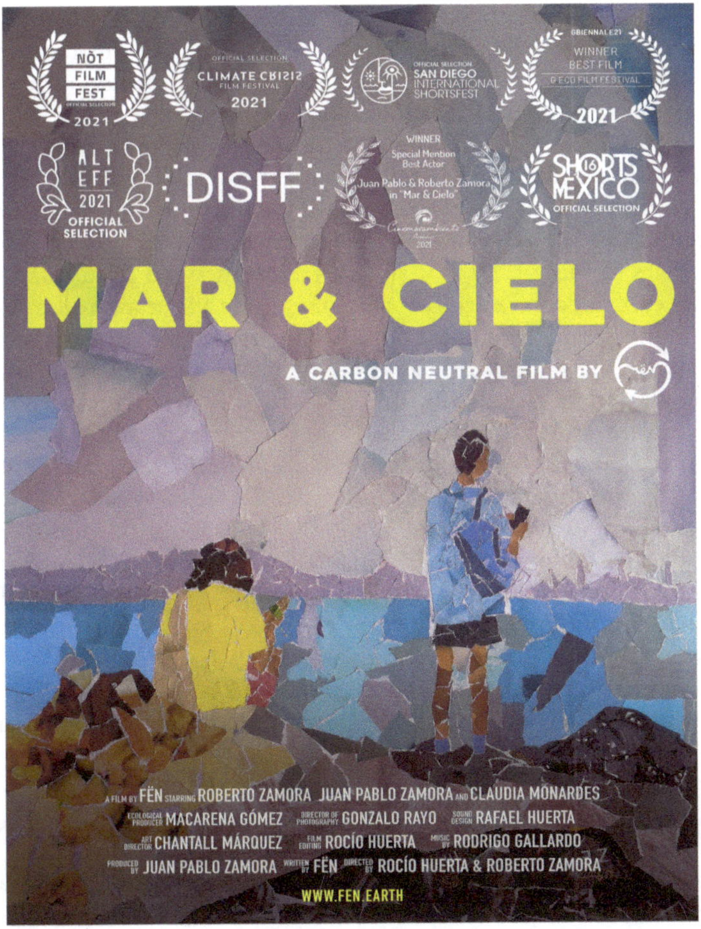

Fig. 16.1 Poster for Mar & Cielo (Zamora, 2021)

Mar & Cielo

By the end of 2020 we produced Mar & Cielo (Zamora, 2021), Chile and Latin America's first carbon-neutral short film (Goldaracena, 2023), recognised in various film festivals around the world such as the Drama International Festival, Shorts Mexico, All Living Things, Not Film Festival and the Cannes Film Market (García, 2021). Mar & Cielo follows two brothers on an emotional journey along the desert coast of Chile, delving into their intricate relationship amidst the backdrop of the climate crisis. Set in Huasco, a town labelled a 'Sacrifice Zone' due to industrial encroachment, the film sheds light on the resilience of its inhabitants and ecosystem.

Given our family's deep ties to the region's history, Mar & Cielo represented our quest to reconcile our past through a form of karmic cleansing. Our paternal grandfather worked at the CAP (Pacific Steel Company), a major contributor to the area's pollution and environmental damage. Despite this, we felt a duty to heal our family history while acknowledging his sacrifices to support us in vastly different times and circumstances.

These experiences fuelled our determination to raise awareness by crafting an authentically ecological film. From the inception of the script, we aimed to address and mitigate any potential emission, we prioritised circular ecologies, meticulously devising a sustainable plan to foresee our carbon footprint. If Mar & Cielo had been made by industry standards it would have emitted approximately 10 tonnes of CO_2, but because of shooting entirely with solar energy, having a zero waste code on set, and vegan local catering for cast and crew, we brought this number down to 0.106 tonnes, which we later mitigated through a tree planting session (Fën, 2023). These calculations were made internally by the production utilising a carbon footprint calculator tool by Albert (2023).

Mar & Cielo was a production that emphasised an optimistic vision of the future and the opportunity to radically change the history of the nation, moving Fën into a more holistic understanding of film production, and planting a seed towards regenerative filmmaking.

Eleven Agreements for Regenerative Filmmaking

Through projects like Mar & Cielo, we recognise our duty as creators to wield our storytelling abilities responsibly, addressing environmental and social issues from inception to distribution. We believe in the transformative potential of narratives, aiming to inspire positive change rather than dwell on dystopian scenarios. Towards regenerative filmmaking, we seek to be an active entity in the transmutation of our society towards collaboration and unity. We believe that a lot of what we see around us is a reflection of our inner being, and therefore we decided to put all our creative energy into the revolution of consciousness and imagination. As a way to hold ourselves accountable to this commitment, whenever we board a new project, we ask ourselves, are we considering our 11 agreements?

Our first agreement is for awareness. Today's reality is not the same as our parents', and climate change is affecting every area of our lives. This crisis impacts policies all over the world as we enter a period fuelled by social and environmental transformation, even affecting our emotions and interpersonal relationships. Our awareness of these changes is our best chance to keep our critical thinking moving towards a kinder and more sustainable tomorrow.

As we move forward into a future driven by environmental innovations and socio-economic transitions, an agreement for sustainability and circularity becomes mandatory. We are aware of the need for an ecological production model that can be sustained over time. Therefore, our production chain is designed to imitate nature's cycles, so everything produced is either recycled or returned to earth.

It is within our agreements to remain committed to sustainable practices and technologies, that we enhance an environmentally sustainable economy, and the various global initiatives that seek to strengthen the management of environmental and social justice. While we adhere to international proposals such as the Escazú Agreement and the UN 2030 Agenda (Soergel et al., 2021), we also propose to go further through innovation within the film industry, to lead the way into a balanced society.

Our fourth agreement for collective collaboration is the way we seek to innovate by building a paradigm where the participation of the collective connects our creative skills with our passion for the environment and social justice. In this, individuals can participate in a horizontal way with leaders instead of bosses. We believe that to heal the planet, it is essential to create networks and alliances that allow people from different sectors and places to come together and chart a common path.

One key moment within a film production that represents coming together is through the food we eat. Through our agreement for food sovereignty (Weiler et al., 2015), we promote the right of every community to eat in a safe, healthy, nutritious, and conscious way. For this reason, our catering commits to be based on responsible production methods that also encourage environmental protection with animal welfare and support the conservation of resources, building a bridge between people, their food and those who produce it. Although there is no academic article to back the following statement, it has been part of our experience when shooting Mar & Cielo, that food tastes better when the cooks sing while preparing it: a performative link to taking care of ourselves through eco-affectivity.

From food comes food waste, from art comes scraps, and overall, from making movies a lot of waste is generated. Therefore, our agreement for a zero waste policy is an essential part of ecological filmmaking. Recycling, reducing, reusing and composting are principles that are never exempt.

Based on our own experience, we are convinced that an audiovisual production can be fully powered by the sun and the wind, hence our seventh agreement with renewable energies. These energies can feed cameras, lights, sound and computers within all stages of a production. In the near future, energy sources such as geothermal, wave energy, and green hydrogen will be more accessible for all involved in our ecological productions (Fig. 16.2).

Since there are so many components of filmmaking that are still largely dependent on fossil fuels, our next agreement is for responsible

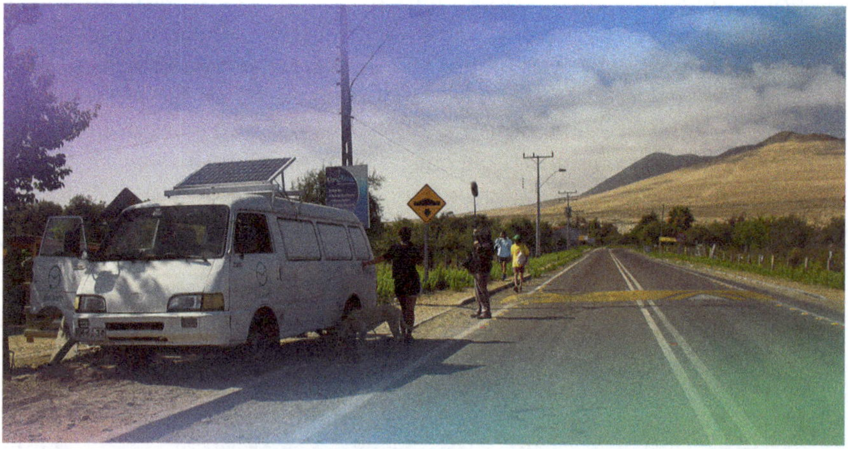

Fig. 16.2 The ecological making of Mar & Cielo, powering set with solar energy (2020)

consumption. We are aware that decisions made about goods and services during our productions have an impact on the environment, so at a large scale the production department is responsible for constantly encouraging sustainable ways for our team to achieve its objectives. Here we pay special attention to the use of fossil fuels for transportation; the proper use of water to avoid scarcity; and the avoidance of polluting chemicals, since these will produce the greatest footprint in a production.

We believe every creator must take responsibility as a co-creator of reality for the impact their stories have on the community and the planet; this is our agreement for responsible co-creation. At a story level we are most interested in dramas that portray hope towards the challenges we face as a society, if we can imagine it, we can create it. Telling stories is part of human nature and it will continue to be so, even when we stop burning fossil fuels.

Our tenth agreement is for eco-affectivity. We recognise and promote the management of our emotional energy in a creative and loving way as it serves to deeply strengthen the quality of our relationships. The care for our emotions must go hand in hand with care for the environment, giving space for each individual to have the possibility of transforming themselves into the best version of themselves. In our productions we practise collective meditation for 10 to 20 minutes before the shooting day starts, followed by a sharing circle in which we honestly ask ourselves 'how we feel' in that moment. These practices gather awareness of ourselves, our colleagues and our environment.

All agreements lead us to the 11th for regenerative filmmaking. Our goal is to produce regenerative films that not only bring relief to the atmosphere but also become part of its healing. In other words, our compensation for CO_2 emissions will always exceed those generated in the production: by acting on the potential of impact investment, meaning to ensure that a part of the budget and profit is invested in climate change mitigating actions, thus creating a secondary impact (Claudelin et al., 2020) via investing in education and sustainable companies.

Now that we understand that the 11 agreements encapsulate guidelines for sustainable film production, we can make an exercise of imagination: what would a day of regenerative filmmaking look like?

As a crew member, you may have arrived on your bicycle to the set today, or perhaps it was raining, so you just got off the tram at our regenerative film studio entrance. From the rooftop of the building, a colleague calls out your name as you rush in. You go up to the greenhouse and the meeting room, you greet the cast and crew as they all begin to sit in a tangled spiral. As you all close your eyes, you are happy you made it in time for the optional morning meditation.

Afterwards, it is time for breakfast, and while coffee brews, you step into the microalgae section of the greenhouse out of curiosity, you have never been there before and you notice how warm it is inside. You almost cannot believe that the 34 degree Celsius heating comes from waste heat generated by the post-production servers from the room below (Schneppenheim, 2020). You laugh to yourself thinking, 'While I am eating a protein packed algae chia pudding from the studio-grown food, someone is creating a 3D environment for this movie on the computers under my feet'. The assistant director calls everyone to set, while one of the green consultants guides a group of extras to the compost bin.

Picking up your reusable water bottle with your name on it from the craft services, right where you left it, you make your way down to the film set. Today you are shooting on the virtual stage (Tangcay, 2022), where solar-powered LED walls (Eveleigh, 2022) surround and illuminate the actors on the scene. The walls are generating a 180 degree view from the centre of an ecological futuristic city. The projection is filled with various elements taken from a realistic utopia such as airborne wind turbines, old buildings surrounded by ivy leaves with blossoming community gardens on their roof, electric trams and extensive bike lanes taking over the old car lanes, new dome constructions built with photovoltaic glass, and a diverse and colourful crowd of people. It is the year 2088 according to the script. The writer-director is obsessed with the concept of solarpunk, an artistic movement that envisions and works towards actualising a sustainable future interconnected with nature and community (Gillam, 2023).

You do your work, may it be as an electrician controlling the spotlight on the protagonist, or as a script supervisor holding an iPad and making sure that on the next shot, the protagonist will point towards the purple bike with his right hand. Or perhaps you are a costume designer, getting ready to clean the mud from the co-protagonist's shoes when he accidentally steps into the wet soil. Whichever job you have on set, it is now your lunch break. So you are sitting in the dining room inside the greenhouse rooftop garden, and you have just finished the vegan lasagna of the day, surrounded by the same family of tomatoes that made the sauce you ate. Once again you pinch yourself to check if you are dreaming.

The second part of the shooting day goes smoothly, and while the cast and crew start to wrap the day, the ecological producer goes around with a list signing people up for the optional weekend activities; they have planned a creativity workshop for children in a nearby non-profit organisation for refugees. You look forward to sharing your knowledge, so you sign up to teach the workshop. Saying goodbye to your colleagues you decide to take your reusable water bottle with you. Tomorrow you will be shooting outdoors at a lake just outside the city using only natural light. The train in that direction has beautiful scenery.

Metaphorically speaking, we believe that days like these can happen by following our 11 agreements in a film production, all the way from awareness to regenerative filmmaking. A circle is created where every completed project can represent a tree, and in which the people who participated in the production are like fruits, therefore seeds, who will spread the agreements in their upcoming commitments. This way, we are confident that through regenerative filmmaking we have a chance to optimise how movies impact both our planet and society as a whole.

Competing Interests 4) The authors have no conflicts of interest to declare that are relevant to the content of this chapter.

References

Albert. (2023). Carbon calculator and production certification by Albert. Retrieved May 13, 2024, from https://wearealbert.org/

Anderson, K. (2023). The six r's of sustainability: What are they?. *Greenly*. Retrieved May 13, 2024, from https://greenly.earth/en-gb/blog/company-guide/the-six-rs-of-sustainability-what-are-they

Antico, F., Wiener, M., Araya-Letelier, G., & Gonzalez, R. (2017). Eco-bricks: A sustainable substitute for construction materials. Revista De La Construcción. Journal of. *Construction, 16*(3), 518–526.

Bailey, M. (2010). *Greenlit [film]*. Ambush Entertainment and Earth Friendly Films.

Becker, C., & Manstetten, R. (2004). Nature as a you: Novalis' philosophical thought and the modern ecological crisis. *Environmental Values, 13*(1), 101–118.

Bilican, F. I. (2016). The relationship between focused attention meditation practice habits, psychological symptoms, and quality of life. *Journal of Religion and Health, 55*(6), 1980–1995.

Carmichael, H. (2023). Think before you print, use less ink. *Care Without Carbon*. Retrieved November 13, 2023, from https://carewithoutcarbon.org/whats-new/events/think-before-you-print-use-less-ink

Claudelin, A., Uusitalo, V., & Hintukainen, I. (2020). Increasing positive climate impact by combining anti-consumption and consumption changes with impact investing. *Sustainable Development, 28*(6), 1689–1701.

Christensen, T. J. (2023). Anika Kruse helps make films green. *Golden Globes Awards*. Retrieved April 17, 2022, from https://goldenglobes.com/articles/anika-kruse-helps-make-films-green/

Duong, T. (2021). Chile's Atacama Desert: Where fast fashion goes to die. *EcoWatch*. Retrieved November 15, 2021, from https://www.ecowatch.com/chile-desert-fast-fashion-2655551898.html

Eveleigh, R. (2022). Solar-powered lights, camera, action! The film studio powered by PV. *Positive News*. Retrieved November 4, 2022, from https://www.positive.news/society/the-film-studio-powered-by-community-energy/

Fën. (2023). *The ecological making of MAR & CIELO (Detrás de Escenas): First look*. Retrieved May 13, 2024, from https://youtu.be/kFqnIHwUIlA?si=Dy29j9WqAfojUBhS

Gaghan, S. (2005). *Syriana [film]*. Participant Media.

García, M. (2021). Chilean cinema displays its diversity and international potential at the Marché du Film. *LatAm Cinema*. Retrieved July 5, 2021, from https://www.latamcinema.com/especiales/el-cine-chileno-despliega-su-diversidad-y-potencial-internacional-en-el-marche-du-film/

Gillam, W. J. (2023). A solarpunk manifesto: Turning imaginary into reality. *Philosophies, 8*(4), 73.

Goldaracena, I. (2023). Ecological cinema: The first carbon neutral short film of Latin America. *Fintualist*. Retrieved April 4, 2023, from https://fintualist.com/chile/cultura/cine-series/cine-ecologico-el-primer-cortometraje-carbono-neutral-de-latinoamerica/

Hoad, P. (2020). Vegan food, recycled tuxedos - and billions of tonnes of CO2: Can Hollywood ever go green? *The Guardian*. Retrieved January 9, 2020, from https://www.theguardian.com/film/2020/jan/09/vegan-food-recycled-tuxedos-and-billions-of-tonnes-of-co2-can-hollywood-ever-go-green

Lebreton, T. (2023). The 3 best alternatives to cobalt mining. *The Eco Experts*. Retrieved June 9, 2023, from https://www.theecoexperts.co.uk/blog/alternatives-to-cobalt-mining

Loreau, M. (2023). *Nature that makes us human: Why we keep destroying nature and how we can stop doing so*. Oxford University Press.

Orlowski, J. (2017). *Chasing coral [film]*. Exposure Labs.

Polakovic, G. (2001). Seeing a greener big screen. *Los Angeles Time*. Retrieved March 27, 2001, from https://www.latimes.com/archives/la-xpm-2001-mar-27-mn-43263-story.html

Ponti, C. (2022). "Vampire energy" is sucking the life out of our planet. *Wired*. Retrieved April 22, 2022, from https://www.wired.com/story/vampire-energy-climate-environment-earth-day/

Rowold, J. (2011). Effects of spiritual well-being on subsequent happiness, psychological well-being, and stress. *Journal of Religion and Health, 50*(4), 950–963.

Scarborough, P., Clark, M., Cobiac, L., Papier, K., Knuppel, A., Lynch, J., Harrington, R., Key, T., & Springmann, M. (2023). Vegans, vegetarians, fish-eaters and meat-eaters in the UK show discrepant environmental impacts. *Nature Food, 4*, 565–574.

Scheidel, A., Del Bene, D., Liu, J., Navas, G., Mingorría, S., Demaria, F., Avila, S., Roy, B., Ertör, I., Temper, L., & Martínez-Alier, J. (2020). Environmental conflicts and defenders: A global overview. *Global Environmental Change, 63*, 102104. https://doi.org/10.1016/j.gloenvcha.2020.102104

Schneppenheim, L. (2020). Windcloud—A green data center in the North of Germany. *Digital Infra Network*. Retrieved November 18, 2020, from https://digitalinfranetwork.com/windcloud-a-green-data-center-in-the-north-of-germany/

Stanton, A. (2008). WALL-E [film] Pixar Animation Studios.

Soergel, B., Kriegler, E., & Weindl, I. (2021). A sustainable development pathway for climate action within the UN 2030 agenda. *Nature Climate Change, 11*, 656–664.

Soler, J., Merce Conangla, M., & Folch, R. (2003). Emotional ecology: The art of positive emotional transformation. *Gestion, 2000.*

Tangcay, J. (2022). How virtual production is helping to cut costs and reduce carbon footprint. *Variety*. Retrieved April 20, 2022, from https://variety.com/2022/artisans/news/virtual-production-small-budget-1235236717/

Warner Brothers (2006). *"Syriana" fights pollution with renewable energy*. Press Release. Retrieved January 18, 2006, from https://www.warnerbros.com/news/press-releases/"syriana"-fights-pollution-renewable-energy

Weiler, A., Hergesheimer, C., Brisbois, B., Wittman, H., Yassi, A., & Spiegel, J. (2015). Food sovereignty, food security and health equity: A meta-narrative mapping exercise. *Health Policy and Planning, 30*(8), 1078–1092.

Winstead, A. (2023). The Art of Sustainable Set Building. *Raindance*. Retrieved September 5, 2023, from https://raindance.org/the-art-of-sustainable-set-building/

Zamora, R., Cifuentes, J., & Jaque, C. (2013). *Film, an ecological activist*. Graduate thesis in Universidad de Artes, Ciencias y Comunicación.

Zamora, R. (2014). *Traffic light [film]*. Fën.

Zamora, R. (2015). *Santiago & Paloma [film]*. Fën.

Zamora, R. (2021). *Mar & Cielo [film]*. Fën.

CHAPTER 17

Diaspora Ecology: A Medicine

Jenn Steverson

Introduction: Mojo Bags

The basis for this chapter is *Mojo for Climate Change*, a project that I started during a fellowship at Winterthur Museum and Library in Delaware, a state in the mid-Atlantic region of the United States. I was inspired by the design of antique seed bags from Dutch Pennsylvania (Winterthur Museum, 2009). The small cloth seed bags reminded me of mojo bags, protective, spiritually charged talismans that are a Black Southern spiritual practice. They are medicines that are imbued with power to help the recipient (Young, 2011; Chireau, 1997). I am creating many mojo bags to honour and support the multitude of small, local adaptation strategies that communities use to survive, connect, and build networks of support (Eicher, 1976). The materials were chosen carefully to create medicine for a just, abundant world where power and resources are shared equitably. The project materials were grown and processed in ways that benefit ecosystems and artisans. The Kala Cotton that is used for this project is an heirloom cotton varietal, grown and woven in Gujarat India. Finimugu is a handspun and handwoven Malian strip cloth made with

J. Steverson (✉)
Geneva Jean Studio and Black Fiber and Textile Network, Central Texas, TX, USA

M. Matthews (ed.), *Ecologies in Practice and Learning: Arts Interventions in the Earth Crisis*, Palgrave Studies in Educational Futures, https://doi.org/10.1007/978-3-031-82338-1_17

sustainably grown heirloom cottons. Both are fairly traded through Botanical Colors, a small business in Seattle committed to ethical sourcing that honours the Earth and workers.

The mojo bags are made with botanical dyes. Colours were chosen based on what grows at Winterthur, and also based on colours that have connections to the African Diaspora. The botanical dyes are herbal medicines: goldenrod, sulfur cosmos, turmeric, pokeberry, and indigo (Barbour & Simmonds, 1971). Goldenrod, turmeric, pokeberry, and indigo are all used medicinally within the African Diaspora. Turmeric is used extensively in the Caribbean where it was brought by indentured Indian immigrants. I purchased the turmeric from a farmer and spice vendor in Grenada, a Windward Isle in The Antilles archipelago. Pokeberry is found throughout the southern United States and is harvested as a food and medicine (Lee, 2014). Indigo blue is a medicinal colour that is used for protection (Parks, 2020).

Context: COVID-19 Pandemic and Inequities of Care

I drew on my experience gardening, and working with community gardens in Austin Texas to select seeds that can thrive in a hotter world. These small portable and personal artifacts connect the global nature of climate change with the intimacy of survival. I conceived of this art installation during the middle of the quarantine period COVID-19 pandemic in the winter of 2021, when a bone deep freeze hit Central Texas. Everything was covered with ice, and then the power started to blink out, and some people lost water as well as heat. Two emergencies overlapped each other, to expose the most vulnerable people and paralyse government response. Community groups and mutual aid—unlikely partnerships—emerged to help. A network of small, localised responses inspired me to think about artisan networks as a solution to the large calamities caused by the textile industry.

Like many others, I needed soothing and decided to put my energy towards a better world. I was reaching for a medicine that could help me through a difficult time when even vaccines were being hoarded by wealthy countries. The pandemic worsened the social fractures in my city Austin, Texas. More people became homeless and hungry. An erratic icy storm took down the state's aging grid and with it went access to fresh water. I have lived through several manmade disasters, from September 11th and its aftermath to the NYC Blackout. They were more frightening in Texas because guns are prevalent and resource hoarding makes each disaster sharper. I live in a state with virtually no safety net. College students struggle to access food and people live in encampments throughout greenbelts, public parks, and under highways. For all its wealth, Austin Texas is desperately short of emergency housing, with years-long waiting lists. Austin Texas has become a hub for technology and media companies. The wealth of Facebook, Tesla, Amazon, Oracle have not contributed to housing equity or trickled down to people who are most in need. In my community, it is clear that technological solutions will not address systemic inequities like racism and poverty.

Limits of Oligarchic Solutions

Many technological strategies are being proposed as a remedy to climate change, and to ecological damage. But most of these solutions lack the heart, creativity, and empathy that is called for in this moment. Technology that relies on exploiting people's labour, displacing indigenous peoples, and devouring resources will not save us.

Too many of the innovative technologies that are meant to build climate resilience have taken on the same old imperialist shapes. Dangerous mineral operations like cobalt mines, fuel electric cars and cell phones, are wreaking havoc on communities and their ecosystems in the Congo (Patisson, 2021).

Elon Musk, among other oligarchs, builds cutting-edge technology for electric cars using exploitative labour practices. The extractive mining practices that are required for the electric car industry are utilising the same exploitative methods as those established by imperialist administrators.

The conditions in Congolese mines are similar to those that Zora Neale Hurston documented in the American South during the Great Depression. In the 1930s Hurston visited turpentine camps during her time collecting songs for the Works Progress Administration. The turpentine workers were confined to camps by armed patrols (Stetson Kennedy, n.d.). Oligarchs like Musk are working to innovate products, but the means of production are the same as those used in extractive imperialist industries.

The old roads of capitalism do not lead to climate resilience or to regenerative systems that heal the earth and our social structures. The people working to change textile production are showing a way forward for other industries. Textile workers have borne a heavy toll due to industrial dangers, from enslaved weavers in antebellum America, immigrants in the Triangle Shirtwaist Factory to the Rana Plaza disaster survivors.

On March 25, 1911, in New York City, the Triangle Shirtwaist Factory caught fire. The exit doors were locked and fire escapes were broken. 146 workers, most of them young women from the Italian and European Jewish immigrant communities died. Some workers, desperate to escape the top floors of the building, jumped out of the windows (OSHA—Occupational Health and Safety Administration, 2011). The Rana Plaza Factory in Bangladesh was the site of a more recent tragedy. In 2013, a nine story garment factory collapsed. 1134 people died (Mahmud, 2023).

The artist Rachel Breen's piece *Shroad* (2018) memorialises the lives lost in both disasters with an assemblage of white cloth that evokes Muslim and Jewish burial rituals. Breen's work also includes installations of deconstructed clothes to demonstrate that all clothing, even fast fashion, is still made by people who are cutting, sewing, and dyeing in factories. During a Fullbright Scholarship in India, she also highlighted the work of RaasLeela, a small textiles company that works exclusively with upcycled and undyed fabrics (Fig. 17.1).

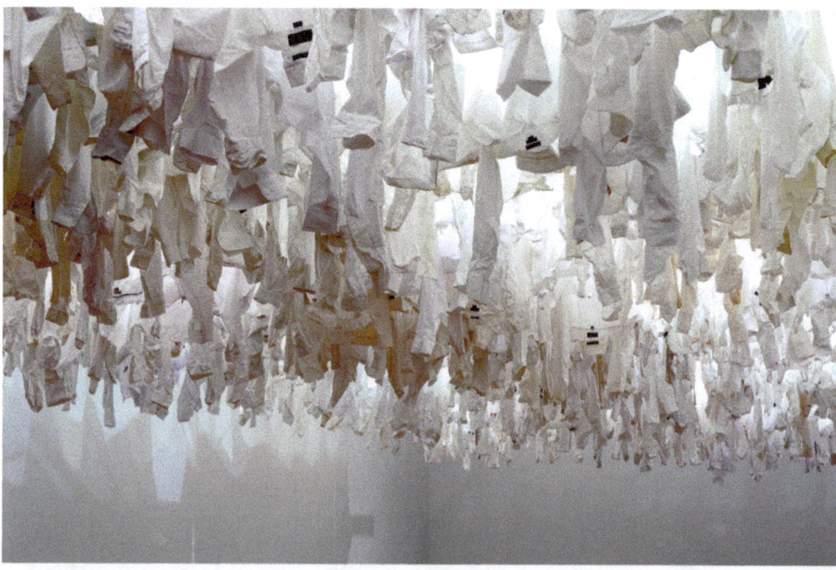

Fig. 17.1 Rachel Breen, Shroad 2018 (image included with the artist's permission)

Craft: Place-Based Climate Adaptation

The continuity and revival of ecologically connected artisan textile networks offer a remedy for climate change, when materials are cultivated in regenerative ways that benefit the workers who weave, dye, and sew cloth. Scientists in the John Innes Centre Germplam Resource Unit (GRU) are studying crop wild relatives of domesticated plants to uncover genetic traits that can strengthen the resilience of food systems (John Innes Research Centre, 2024). Their work is focused on the key food crops in the United Kingdom. Researchers have also studied older varieties of cotton, like kala, a rain fed cotton that is indigenous to India (Jha, 2018).

There is an open loop of communication between the academic research, farmers, and processors. The GRU research is made possible by a seed collection that dates to the 1920s when there was more genetic diversity in food crops. The study of kala cotton examines the use of this heirloom fibre by fashion designers and textile mills. Solutions are being created by small companies and cooperatives that are creating models built on fair wages, waste reduction, and ecological responsiveness. Governments and non-governmental organisations need to look to these grassroots organisations as a model for how to make the changes that are needed to mitigate climate change.

More scholars are beginning to speculate about economic degrowth as a resilience strategy. The World Economic Forum broadly defines economic degrowth as 'shrinking rather than growing the world's economies to use less of the world's dwindling resources' (Masterson, 2022). Degrowth in the textile industry could help to address issues like waste colonialism. This concept was first developed in 1989 at the Basel Convention on the Control of Transboundary Movements of Hazardous Waste and their Disposal to describe the transport of toxic waste from wealthy nations to developing nations (United Nations Environmental Program and Secretariat of the Basel Convention, 2020). The Or Foundation facilitates the Stop Waste Colonialism Project to address the impact of cheap textiles on second-hand market sellers and their communities in Ghana (Rickets & Skinner, 2023).

Stop Waste Colonialism calls for the companies that profit from fast fashion to adopt Extended Producer Responsibility (EPR). This is defined as 'an environmental policy and form of product stewardship that extends a producer's responsibility for a product to the post-consumer stage of a product's life-cycle' (Rickets & Skinner, 2023, p. 1). I consider this to be part of degrowth because it is a reconsideration of how companies account for their costs and profits.

The fast fashion industry is creating a glut of cheap clothes that are flooding second-hand markets in Ghana with poor quality textiles (Verges, 2023). These materials end up in landfills, beaches, and waterways (Masterson, 2022; Rickets & Skinner, 2023). Slow could be the right speed for ecological and humanistic regeneration (Adisa-Farrar, 2022).

Enslaved African people created goods on a mass scale for cottage industries that were on the precipice of industrialisation. Their labour made industrialisation possible and it was fuelled by the disconnection between raw materials produced in one place, processed in another, and wealth concentrated by the first transnational corporations. I recently visited an old cotton mill, and saw indigo pits in Carriacou, which is part of Grenada. This site connects the period of enslavement with the contemporary calls for reparations that have been issued by the Caribbean Community (CARICOM), an intergovernmental association of 20 members. Mia Mottley, the Prime Minister of Barbados, is a leading strategist in linking reparations with climate justice (Lustgarten, 2022). Colonial powers extracted labour and raw goods from the territories that they dominated.

In the Caribbean, British enslavers were even compensated for lost labour after emancipation (Anson & Bennett, 2022). Reparations are needed to repair this harm, and to help a region that has is being heavily impacted to become more resilient. Climate resilience can be achieved through infrastructure projects, but there is also a transition that could be made to the agricultural sector, to help small island nations achieve food sovereignty. Food sovereignty is deeply connected to seeds, and helping smallhold farmers to adapt to changing growing conditions (Altieri & Koohafkan, 2008). The mojo bags I make relate to this intention to highlight the role of heirloom seeds in food sovereignty and climate resilience.

Mojo Bag Cloth

The seventy seven mojo bags I have made are filled with heirloom seeds that have been adapted to thrive in hot and dry climates (Fig. 17.2). In addition, there are seeds that are offerings for historically enslaved Africans (Carney & Rosomoff, 2009). Foods like Sea Island red peas, benne, sorghum, and hominy are included in the bags. The cloth that was used for this project is a combination of upcycled cloth, handwoven kala cotton indigenous to India, and finimugu—a handspun Malian strip cloth for the project (Botanical Colors, 2024). Both are fairly traded to benefit the artisans and farmers who grow and process the plants. Crops for the cloth are grown in ways that nourish the land. The growing conditions, weaving process, and worker compensation were an important consideration

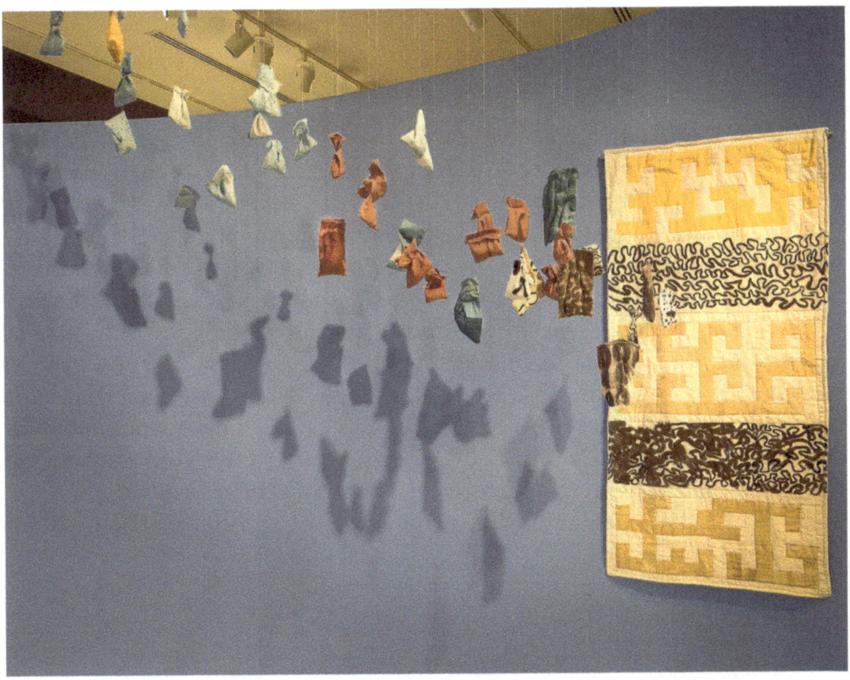

Fig. 17.2 Jenn Steverson, Mojo for Climate Change. In, *Transformations: Contemporary Artists at Winterthur*, 2024–25, (detail)

for this project because it is a response to climate change, an issue that requires an intersectional response that transforms living conditions for all beings.

The way that cotton is handled influences its texture and weight. The texture and weight of fabric is called 'the hand' of the cloth. This is used to describe its roughness, or smoothness, whether it is flatly woven or retains more texture. When cotton is hand-spun, there are variations in the thickness of the thread that are visible in the final cloth. The variations and changes in the hand of naturally treated cotton are unique and beautiful. Each skein of thread, each yard of fabric is unique. This uniqueness is also a signature of sorts. Textile scholars can examine cloth to learn about the type of looms used to weave the cloth, the way that raw fibres were processed through spinning, washing, and dyeing (Harris, 1993). All of this richness is part of the mojo bags.

Herbal Colours: Goldenrod, Pokeberry, Indigo and Turmeric

The colours are a medicine. Mojo bags were often made from red flannel (Young, 2011), and the colour blue is used as protection in Gullah/Geechee culture (Shoshi Parks, 2020). Gullah Geechee are African descendants who maintained language and craft practices due to their relative isolation during the period of enslavement. The Gullah Geechee historically lived in coastal areas and islands that reach from North Carolina to Florida. Blue beads, blue bottles, and blue paint anoint the entrances of homes and yards. There is a magic to tending an indigo vat: a peace settles over me when my hands are dyed blue to the wrists, my fingernails and cuticles holding on to the colour for the weeks that I am working my vat. The yellow green goldenrod dye was made using flower spikes that were harvested on the Winterthur grounds. The berries have been used medicinally by Black Southerners for generations (Lee, 2014; Allen, 2015) (Figs. 17.3 and 17.4).

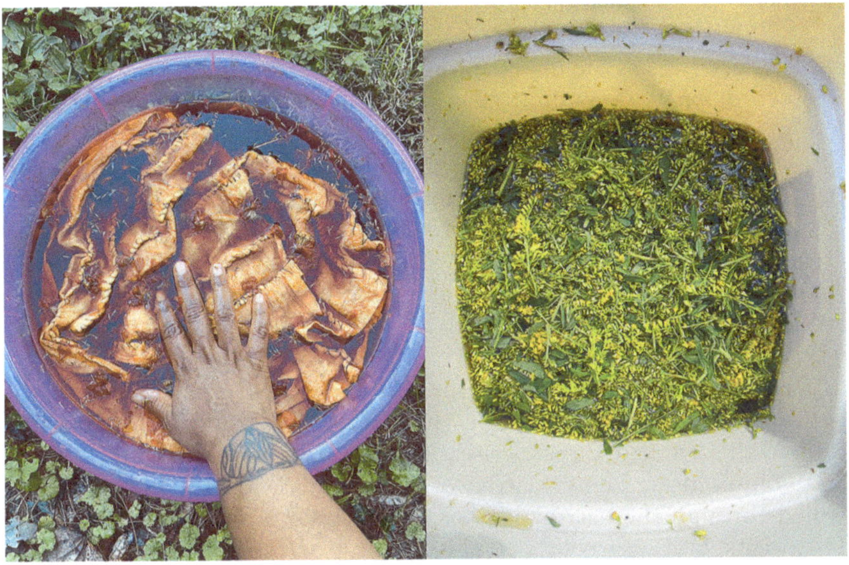

Fig. 17.3 Tending to the indigo vat and goldenrod

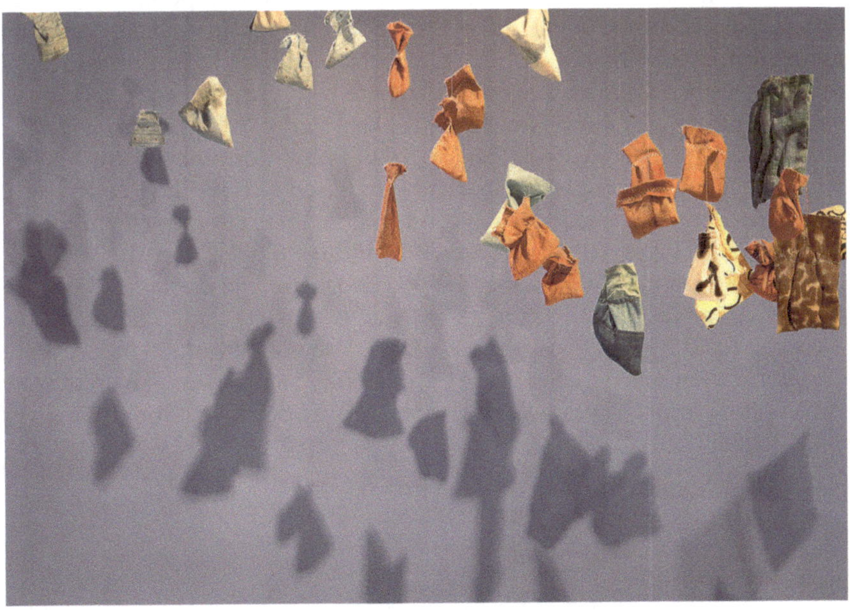

Fig. 17.4 Jenn Steverson, Mojo for Climate Change (detail)

Conclusion

My practice lives at the intersection of art and research. I comb archives for records of Black communities, with a special attention to placemaking practices, textile crafts, gardening, and spiritual practices. I sift through materials that are often tainted with white supremacist ideology, records steeped in assumptions about racial superiority, and primitive peoples. Historic archives with records about Black lives are rich resources collected by untrustworthy narratives. Reading primary sources from the eighteenth, nineteenth, and twentieth centuries requires sorting opinion from fact, and noticing silences as much as words (Hartman, 2019).

Delving into archives about the Black Diaspora that document the period of enslavement brings to the surface the economic drivers that created that current climate crisis, through forced labour, displacement, and colonisation. The valuation of land, materials, and peoples is funnelled through the narrow lens of cash, capital, investment, and profit. Historians and journalists are beginning to tell this story of climate change.

The immense scale of climate change is overwhelming. The problems unfolding in the world are rooted and driven by the imperialist shape of the global economy that controls the concentration of wealth and the extraction of raw materials. Black feminist scholars have illuminated a different view of power that is an antidote to the concentrated, centralised models. Thinkers like bell hooks, Saidiya Hartman, and the Combahee River Collective (Taylor, 2017) have illuminated the ways in which Black women specifically have organised and worked collectively to shape their communities. Bayo Akomolafe, a contemporary philosopher, describes the Yoruba notion of building power, and the way this generated a parallel world that helped Black people to survive on the other side of the ocean during the Atlantic Slave Trade (Hemphill & Akomolafe, 2021). The path forward must be polyphonic to incorporate voices that have been silenced. People at the margins are shaping new visions for a more resilient world and generating power for themselves (Fig. 17.5).

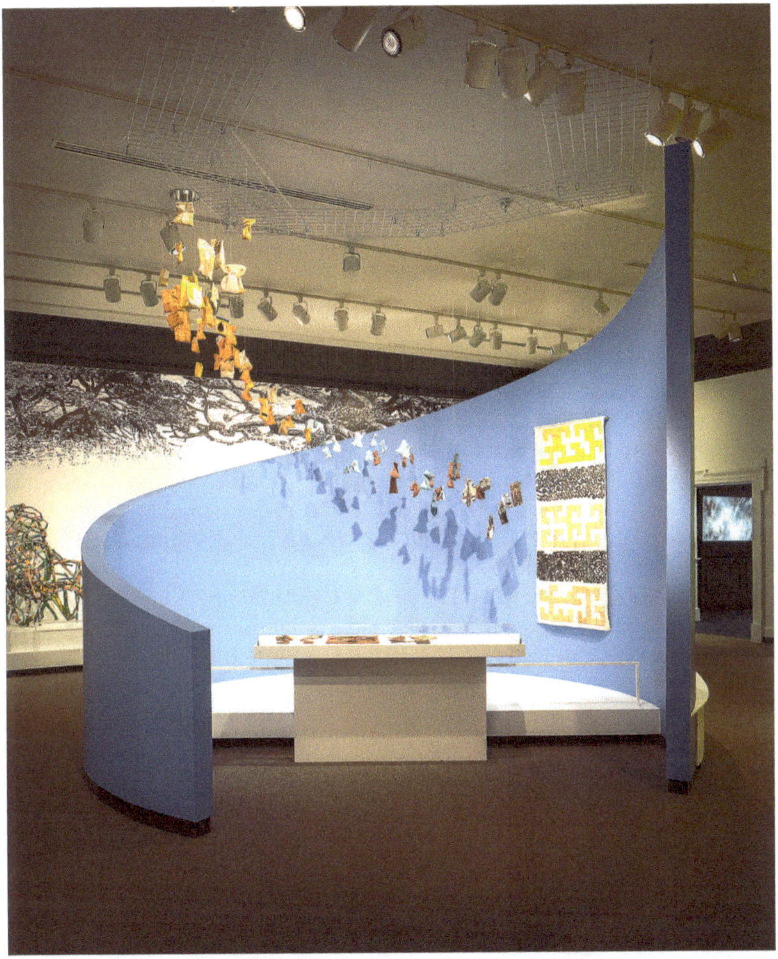

Fig. 17.5 Jenn Steverson, Mojo for Climate Change, 2024–25

Competing Interests Jennifer Steverson has received a project grant from Dumbarton Oaks and a Maker-Creator Fellowship from Winterthur Museum and Library.

References

Adisa-Farrar, T. (2022, January 12). *Regenerative design: Textile systems.* Retrieved July 15, 2024, from https://slowfactory.earth/courses/regenerative-design/

Allen, R. (2015). A southern recipe for natural ink. *New York Times Style Magazine.* Retrieved July 15, 2024, from https://www.nytimes.com/2015/09/17/t--magazine/how-to-make-ink-from-pokeberry-fruit.html

Altieri, M. A., & Koohafkan, P. (2008). *Enduring farms: Climate change, smallholders and traditional farming communities.* Third World Network.

Anson, M., & Bennett, M. D. (2022). *The collection of slavery compensation, 1835–43* [Working Paper]. Bank of England. Retrieved July 15, 2024, from https://www.bankofengland.co.uk/working-paper/2022/the-collection-of-slavery-compensation-1835-43#:~:text=As%20part%20of%20the%20compromise,behalf%20of%20the%20British%20government

Barbour, J., & Simmonds, D. (Eds.). (1971). *Adire cloth in Nigeria.* The Institute of African Studies: University of Ibadan, Tater Blue Library.

Botanical Colors. (2024, March 1). *Finimugu (Malian strip cloth)* [Commerce and education focused on botanical dyes and cloth]. Botanical Colors. Retrieved July 15, 2024, from https://botanicalcolors.com/shop/fabrics-and-dyeables/finimugu-malian-strip-cloth/

Carney, J. A., & Rosomoff, R. N. (2009). *In the shadow of slavery: Africa's botanical legacy in the Atlantic world.* University of California Press.

Chireau, Y. (1997). Conjure and Christianity in the 19th century: Religious elements in African American magic. *Religion and American Culture: A Journal of Interpretation, 7*(2), 225–246.

de Verges, M.. (2023, May 23). *Ghana's battle against 'waste colonialism' in the wake of fast fashion.* Retrieved July 15, 2024, from *Le Monde.* https://www.lemonde.fr/en/environment/article/2023/05/23/in-ghana-the-battle-against-waste-colonialism-in-the-wake-of-fast-fashion_6027668_114.html

Douglas, E., & Schumacher, Y. (2023, September 7). Texas just recorded its second hottest summer on record. *Texas Tribune.* Retrieved July 15, 2024, from https://www.texastribune.org/2023/09/07/texas-hottest-summer-2023/

Eicher, J. B. (1976). *Nigerian handcrafted textiles.* University of Ife Press.

Harris, J. (Ed.). (1993). *5,000 years of textiles.* Smithsonian Books.

Hartman, S. V. (2019). *Wayward lives, beautiful experiments: Intimate histories of social upheaval.* W.W. Norton & Company.

Hayhoe, K. (2021, November 11). A climate for all of us. *The Nature Conservancy,* Winter 2021. Retrieved July 15, 2024, from https://www.nature.org/en-us/magazine/magazine-articles/climate-justice-hayhoe/

Hemphill, P., & Akomolafe, B. (2021). *Hope, questioning and getting lost (3).* Retrieved July 15, 2024, from https://podcasts.apple.com/us/podcast/finding-our-way/id1519965068?i=1000519932046

Jha, B. (2018). *Kala Cotton: A sustainable alternative. Surviving and thriving.* The Asian conference on sustainability, energy and the environment, Kobe. Retrieved July 15, 2024, from https://papers.iafor.org/proceedings/conference-proceedings-acsee2018/

John Innes Research Centre. (2024). *Germplasm resource unit [Academic].* John Innes Centre. Retrieved July 15, 2024, from https://www.jic.ac.uk/research-impact/germplasm-resource-unit/

Lee, M. E. (2014). Working the roots: Over 400 years of traditional African American healing. In J. Douglas Allen-Taylor (Ed.). Wadastick.

Lustgarten, A. (2022, July 27). Barbados resists climate colonialism in an effort to survive the costs of global warming. *ProPublica and New York Times.* Retrieved August 7, 2024, from https://www.propublica.org/article/mia-mottley-barbados-imf-climate-change

Mahmud, F. (2023, April 24). Ten years of Rana Plaza: How safe is Bangladesh garment industry? *Al Jazeera.* Retrieved July 15, 2024, from https://www.aljazeera.com/news/2023/4/24/ten-years-of-rana-plaza-how-safe-is-bangladesh-garment-industry

Marya, R., & Patel, R. (2021). *Inflamed: Deep medicine and the anatomy of injustice.* Farrar, Straus and Giroux.

Masterson, V. (2022, June 15). *Degrowth - What'S behind this economic theory and why does it matter right now? [NGO think tank].* World Economic Forum. Retrieved July 15, 2024, from https://www.weforum.org/agenda/2022/06/what-is-degrowth-economics-climate-change/

OSHA (Occupational Health and Safety Administration). (2011). *The Triangle Shirtwaist Factory Fire.* (2011, March 25). United States Department of Labor. https://www.osha.gov/aboutosha/40-years/trianglefactoryfire

Parks, S. (2020, January 14). What the color 'haint blue' means to descendants of enslaved Africans: In the Lowcountry, the unique shade is both a protective talisman and source of unspeakable suffering. *Atlas Obscura.* Retrieved July 15, 2024, from https://www.atlasobscura.com/articles/what-haint-blue-means-to-descendants-enslaved-africans

Patel, R., & Moore, J. W. (2017). *A history of the world in seven cheap things: A guide to capitalism, nature, and the future of the planet.* University of California Press.

Patisson, P. (2021, November 8). 'Like slave and master': DRC miners toil for 30p an hour to fuel electric cars. *The Guardian.* Retrieved July 15, 2024, from https://www.theguardian.com/global-development/2021/nov/08/cobalt-drc-miners-toil-for-30p-an-hour-to-fuel-electric-cars

Rickets, L., & Skinner, B. (2023). Stop waste colonialism: Leveraging extended producer responsibility to catalyze a justice-led circular textiles economy. *The Or Foundation.* https://stopwastecolonialism.org/stopwastecolonialism.pdf

Stetson Kennedy, W. (n.d.). Mark of Zora. *Library of Congress, Works Progress Administration*(Florida Folklife from the Works Progress Administration Collections, 1937-1942). Library of Congress. Retrieved July 15, 2024, from https://www.loc.gov/collections/florida-folklife-from-the-works-progress-administration/articles-and-essays/a-florida-treasure-hunt/mark-of-zora/

Taylor, K. Y. (Ed.). (2017). *How we get free: Black feminism and the Combahee River collective*. Haymarket Books.

Thompson, K., & Reader, G. (2022, February 14). A year later: Looking back at Texas 2021' winter storm. *KXAN*. Retrieved July 15, 2024, from https://www.kxan.com/news/texas/a-year-later-looking-back-at-texas-february-2021-winter-storm/

United Nations Environmental Program & Secretariat of the Basel Convention. (2020, April). *Basel convention: Controlling transboundary movement of hazardous waste and their disposal*. United Nations Environmental Program. Retrieved July 15, 2024, from https://www.basel.int/TheConvention/Overview/tabid/1271/Default.aspx

Winterthur Museum. (2009). Cotton seed bags, c. 19th century, Winterthur Museum Collection.

Young, J. R. (2011). *Rituals of resistance: African Atlantic religion in Kongo and the Lowcountry South in the era of slavery*. Louisiana State University Press.

PART VI

Ecosophy and Aesthetic Imaginaries

This final section of the ecologies in practice and learning collection turns to voices that are known for theorising creative practice and ecologies. The authors in this section contribute to fields in the arts, education and philosophy. Although each author takes up a different approach in relation to learning ecologies, they can be viewed as developing new 'ecosophies' that imagine outside the current expectations for arts practice, education and global cultures.

Erin Manning moves beyond landscaped parameters for learning in arts practice and academic conventions for how the narratives of practice research can be presented. Manning presents an ecosophical inter-woven account of her time working in the forest, in the Canadian practice of collecting maple syrup—'mapleing'. This forest-time co-exists alongside Manning's development of a body of artwork for an exhibition titled '100 Acres'. Manning presents the living words of her creative field notes. She also enters into a theorisation of 'distantism' as an unsettling mediation of the critical distances created around artworks, the sensitive time cartographies of the mapleing process, and the complexities of material immediacy in the forest.

Chris Turner combines interpretations of ecology with aesthetic theory in arts education to form 'aesthoecology'. With this positioning, Turner offers critical tools to view from the present of 'anthropogenic climate change' into the future. Turner argues that an understanding of aesthoecology—its affects and sensitivities—is crucial to developing arts pedagogies that respond to the Earth Crisis. The visualised alternatives that

Turner offers here are viewed as supporting a more pacificist, less extractive, life-art continuum. Turner gives an account of how the entangled, material processes of arts practice 'come alive', through ecological aesthetics that offer sensory mediations among humans, other-than-humans and the more-than-human.

Dennis Atkinson theorises 'cosmopedagogies' for a view of learning processes that can happen in the cosmos, without standardising space and time relations, or superimposed 'templates' for education. This approach situates our earthed places of study within multiple possibilities and divergent cosmic dimensions. Atkinson discusses how possibilities can be realised in pluralist arts 'instaurations' that can, even in times of crisis, install the potential for different futures to happen. Atkinson argues that such cosmicities need to be made available for the marginalised, such as the disabled and neurodiverse, in compassionate learning in the arts that can free perceptions of capacity—and re-wild learning pathways.

In these evocative materialisations of thought, affect and making, the writings of Erin Manning, Chris Turner and Dennis Atkinson resonate with constellations of ideas arising in the previous sections; for example, in the aftermath of conditioning neoliberal performativity, ecosophy situates value for the resourcefulness of inquiries emerging in practice research. Care for human and other-than-human life forms and mindful stewarding of resources creates harmonics with research presented earlier that takes posthumanist and ecological theory into educational settings in schools, universities, museums, galleries and communities. Manning, Turner and Atkinson interweave the material, spatial and empathic, with the beyond-structures of aesthetics in the process of becoming.

CHAPTER 18

100 Acres

Erin Manning

Introduction

100 Acres, the artwork (Richard Saltoun Gallery, London May 30–June 22, 2024), was made between November 2023 and March 2024, tufted, sewn, woven, knotted, in my studio in Montreal, three hours south of the 3Ecologies Project. Moving from forest to studio, for the first time in 3 years, was an experiment in itself: could the more enclosed space of the studio generate potential in ways congruent with the larger scale of the living studio of 3E?

The work had to be finished by March in order to turn my attention to the mapleing. 3E has a sugar shack on an old-growth maple forest. We have 800 tapped trees. Our small syrup economy offers a living practice of alchemy, teaching us how to move with the weather, the forest, the fire. It's a startling process, intense, exhausting, beautiful. Many people come through to assist, and each batch is its own surprise, taste and colour affected by the atmospheres.

E. Manning (✉)
Concordia University, Montreal, QC, Canada

M. Matthews (ed.), *Ecologies in Practice and Learning: Arts Interventions in the Earth Crisis*, Palgrave Studies in Educational Futures, https://doi.org/10.1007/978-3-031-82338-1_18

Work happens in an acrossness of tendency. 100 Acres, the land, bleeds into 100 Acres, the artwork. The protocol for 100 Acres, the artwork, printed at the end of this piece, leads us back into the forest, on its terms.

+++

Maple syrup is made from the sap of sugar maples. The maples produce sap year-round, but under precise conditions, their sap travels back and forth from their roots to their branches and can be harvested. This happens when the nights are below 0 and the days above 5 degrees Celsius. But this is not a precise science—the trees are sensitive to wind, to sun, and to threshold climate variations. Sometimes the sap will flow all night even though the temperature is cooled to almost freezing, and sometimes the temperature will be above 0 and the trees will hold onto the sap in their branches. Lore says you know the sap is running if your gutters are dripping with melting snow and ice.

I start working on 100 Acres in November 2023. It's been three years since I was last in the studio. During that time, the 3Ecologies Project has monopolised my energies. Days which might have been given to entering the studio to feel into the limits of a textural problem have been taken over by a canvas that feels at once impossibly vast and resolutely pragmatic. The pragmatism of demands oriented by weather conditions and off-grid life have muted that speculative angle of my existence, moving me, daily, into those kinds of tasks that are bounded by necessity. But even when shovelling snow off the solar panels, or cording wood, the feel of the studio is with me. I love being pulled into activation by the urgent necessity of a life that depends on certain parameters being met. Still, I am aware that this hands-on work also relies on the other kind of work, that doing that is resolutely useless, that grows from the crafting of a generative lure, not towards the completion of a task but towards the generative force of an activation contour through which life catches itself obliquely. In angular perspective, art's force is that quality of inflection it torsions, that mode of existence it agitates.

Our maple forest is an old-growth forest, some trees as old as 400 years. The old, craggy trees, sometimes a meter around, aren't the best sap-trees. A more streamlined operation would cut them down in lieu of younger, more vital specimens. They also wouldn't leave as many fallen trees on the ground. Or leave such a range of trees—birch, beach, pine, fir. They would clean things up. Make roads for an ATV or a tractor. We don't do any of that. We are interested in the sap, but also uneasy when the conversation turns extractive. How could it not: the economy of maple syrup is not viable otherwise. To profit, it's necessary to accelerate the potential of the tree's gift: you 'vacuum' the sap, pulling out more than the tree gives of its own accord; you set up an 'osmosis' system which separates a percentage of water from the sap so that you cook a sap that has more sugar content (sap comes out between 1.5–4 percent sugar, and needs to be boiled down 40 times to make syrup); you set up automated systems that take the operation out of your hands.

The return to the studio feels uneasy at first. How to move from the pull of the everyday—that very pragmatic orientation towards what needs to be done in that more abstract world of an artistic space that must create its own conditions, that must generate work from a series of constraints that are—to some degree at least—distanced from the world?

We make syrup the traditional way. The earliest accounts of making maple syrup are indigenous. Fires would have been tended, probably outdoors, boiling down the sap in cauldrons over hours and days. Since colonisation, this practice has expanded in the settler-colonial population, taking over that geography of Canada and the US where the necessary spring freezing-thawing window still exists. In the area of the 3Ecologies Project, three hours north of Montreal, there are many sugar shacks, most of them larger and more automated than ours. As climate change shifts the recognisability of our seasons, and wildfires burn down remaining maple forests on the edge of the boreal forest, there is a sense that the window might be closing. But for now, we boil, making a bit of pocket change that is mostly eaten up by what we unkindly think of as the syrup mafia—that nationalist bank that hoards syrup like gold, imposing restrictions on its harvest. To sell to the mafia is to accept that your syrup will be mixed with others of similar light-transparency (gold, amber, dark), effectively losing that atmospheric singularity that distinguished it not only from forest to forest but from batch to batch.

I begin the work in November because I know I have to be done before the sap starts to run in March. When the sap runs, there is nothing else: the mornings begin with lighting the two fires (the large evaporator that boils the sap and the smaller wood stove that heats the cabin and keeps the water warm for the cleaning we will do throughout the day), taking out the axe and splitting some hard and softwood, bringing in the wood for the first few fires, cleaning and setting up. It takes at least 3 hours until the first batch. We think of it in terms of fires: some batches have been made in 5 fires, but that's rare. In the beginning you make the fire every 30–45 minutes. As the day progresses, sometimes you make it every 15 minutes. It depends on the wood, and on the weather, and on the sap. But we are not there yet, and so there is time to turn attention elsewhere.

100 Acres is made by pulling monks cloth over my wood frame from a 25-yard roll attached above. There are two 25-yard rolls—the finished work will be 50 yards. The frame is 5 feet by 5 feet. I use hand tools and machines: 2 tufting guns, one that loops and one that cuts, 3 hand tufting tools, and a knotting tool. There is a loom in the space that I weave small pieces on, adding them sporadically to panels. I sometimes hand sew directly on the frame, and every once in a while, I add pieces of cotton batting. But for the most part, 100 Acres is tufted. The only difference from a tufted rug is that I work on both sides, interested in troubling the idea of frontality, and the distantism that accompanies it.

Distantism, as John Lee Clark writes, has to do with the belief that the world comes to us at a distance, and that its value lies there, in that capacity to take it in through those modalities of distance we most rely on: sight and hearing (Clark, 2023). Clark is DeafBlind, a leader in the ProTactile movement that teaches us that distantism is a value-laden habit that debases forms of encounter with the world that happen in the amodality—what Clark would call the metatactility—of experience felt, of worlds in the touching.

I tell myself I will 'walk the lines' every day. The lines are the tubing from tree to tree that connects the trees in a meandering spiderweb, sap running through them when the trees are tapped in early spring. While the most traditional way to harvest sap is with buckets, many of us use tubes today: it's much less labour intensive to have the sap gather into one central vat than it is to have to pick it up at each tree. Especially when you have 800 tapped trees, as we do, and no easy path for a vehicle.

Walking the lines is necessary—it ensures that no tree has fallen on the fragile tubing, causing leaks and breakage. One of the first things you learn when you buy a hundred acres of forest is that trees fall all the time. What had seemed to me a relatively solid landscape is actually mobile and shifting. Forests change continuously. It's a remarkable lesson, to realise how fragile these giant trees can be, especially in times of climate change where weather conditions shift so rapidly, where periods of drought are followed by too much rain, or unusual wind storms that, under more predictable conditions, a healthy tree might be able to overcome, but a stressed tree can't survive.

I set up two constraints for the shift that occurs every time I bring down monks cloth, from above, having finished a panel: (1) with consistency, diverge; (2) add a new colour. The proposition: use it all, including the remains. Compost wherever possible.

The first frame is the only time I will work with a continuous open surface. After that, I will always be working into the work—bringing down the monks cloth halfway, building on what is already there. The temptation, given my temperament, will be to hold to a relatively minimalist continuity of line and movement. The first constraint challenges this: what is added cannot simply conform to what is already there. New movements are necessary. Adding a colour creates contrast, challenging my desire for the monochromatic. With contrast, new gestures become possible. Colour turns the work, tweaks it, undoes its pull to self-continuity. It's a challenge to strike the balance between continuity and difference: I often spend an hour in the morning sitting with the work before making a mark. It's hard to inflect into a continuity, attuning to consistency while practicing difference. I follow more than I lead. Soon, it draws me in, becomes a drawing machine.

When we purchased the hundred acres and began the 3Ecologies Project, we soon realised we had to cut trees. Solar panels don't work well in the middle of a forest. Brian and I were uneasy. We didn't want to cut into the land, to be in the settler-colonial habit of settling through clearing. We were conservative with the cut, and we felt it come November, when for weeks there was barely any electricity from solar power. Over time we realised that we saw the trees apart from the forest—that we couldn't yet see the forest as a living ecology. Some culling isn't settling—a clearing isn't a clear-cutting. Locals taught us this, lumberjacks who could show us, as our friend Rémi did, that cutting a tree can promote an adjacent ecology. We had been too precious about the trees themselves, too focused on them. Still, we weren't completely wrong. Had we listened to others regarding the maple forest, we would have cut too deeply into its complex ecosystem. It's better to cut less, and live with the emergent ecology, than enter with a disregard for life, as most often happens. In time, we cut more trees—are even cutting them as I write—hoping to have enough sun in November and December to avoid using the generator and the fossil fuels that run it. We are still learning how to balance the strategic weeding of a local environment, and the more indiscriminate gesture of clearing to extract from the land its capacity to generate its relational webs. We know we are treading a fine line. In the uneasiness of this tension, towards the end of the first year, I write Out of the Clear (Manning, 2023) to work through some of these paradoxes, haunted as I am by the cleared images of residential schools and the graves of indigenous children that are being found in their barren yards.

When the first 25 yards of 100 Acres come off the frame, I wonder how it will weather the passage from the tautness I've been working with towards the more open weave of its natural folds. The weight is immediately apparent, the textile thickening around dense areas of yarn, gaping where the weave remains unworked. There is an animality to it, an uneasy textural proximity. Distances fall away (Fig. 18.1).

I walk the lines. Large downed trees, their roots in the air, have started to grow moss, mushrooms also beginning to appear now that the snow is melting. The sugaring-off feels like a parenthesis with winter on one end and spring on the other—the four to eight weeks of sugaring off marking the shift. Often too busy in the sugar shack to really notice the transitions, there remains the feeling that with every day of sap, winter is lagging

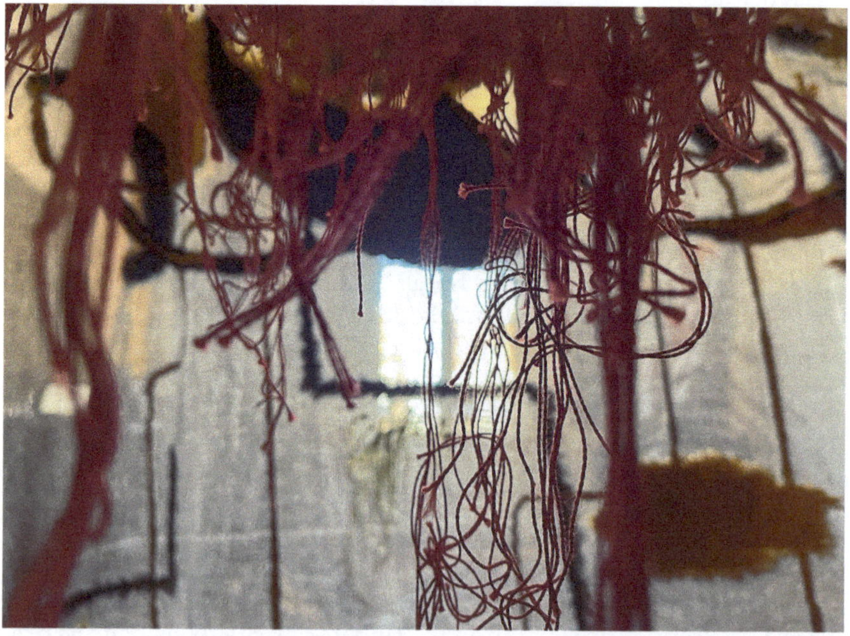

Fig. 18.1 100 Acres, Erin Manning

further behind, despite the nights remaining cold and the forest floor dappled with snow. What we will be watching for, in these weeks of boiling sap, is the presence of buds. When the buds come out, the sap will have turned milky, and mapleing will be over. It's an abrupt change: first, just as the buds are beginning to emerge, the sap will become prolific, though its sugar-content will lower. In a year like the last one when there was almost no sap, this will feel like a bounty. But, as I learned then, that flow will be the marker of a shift. In the boiling, there will no longer be a smell of maple—instead an earthy smell will dominate and the very dark syrup will taste more like caramel than maple. And then, often by the next day, the sap will become cloudy and start to stink. You don't want to still be cooking by then, but if you make a mistake, you will know it.

As I work on 100 Acres, I think a lot about distantism, about the practice of keeping art at arm's length, in the ocularcentric space of a 5-foot abeyance. We look from afar, scanning in the establishing-shot practice of knowing at a distance. And from there we judge, our human gaze perpetuating a mode of existence that takes its centrality for granted. This settler-gazing, this whiteness, is rarely ascribed to the distantism art environments provoke. Thinking with ProTactile makes other ways feel possible.

The forest is not distantist. Everything is right there, confounding compass. Even for those people who navigate it well, it requires landmarks. Landmarks have to be generated in the event, though. To assume you already know can easily leave you walking in circles, completely turned around.

In the maple forest, the web of tubing affords a certain distantism. You can keep your bearings by following the lines—two black mains going up in a v, blue side lines connected to them, the spidery veins that move from tree to tree. When it runs, the sap moves from the blue lines into the mains, which carry it down the hill to the 400 gallon vat behind the sugar shack, the direct conduit to the evaporator where we boil the sap. Walking the lines involves tracking the mains and their tentacular reach, making sure everything looks taut. You make sure the sap is flowing and watch for trees that may have fallen on the lines. If it isn't flowing, you find the chewed-through tubing and you fix it. Or you cut the tree that fell on the line.

But I don't actually get to walk the lines as I had hoped to. It's just too time-consuming to cook, and this year—unlike the last—the sap flows freely and abundantly. We repaired the lines during the tapping late

February, so I am pretty confident the forest looks good. Still, to be mostly focused on the evaporator is to miss the process at tree-level. You need a team, and this year we're short. The sap has started to flow a month early, getting in the way of people's schedules. Climate change is not only about a heating planet—it's about the unpredictability of continuously changing conditions. I feel the unpredictability keenly as we head into our seventh week of 12-hour days, 10 cords of wood burned and 93 gallons of syrup made.

A few years ago, I made some rugs for the DeafBlind community. John and I were talking about architectural thresholds in houses, about the way adjacencies might be better felt. John had told me about a rug he had in his bathroom, and how one day, the rug having been displaced, everything felt just a bit off. The rugs I made became one way for us to think about the presuppositions of distantism not only in architecture, but in hearing-seeing spaces. More touch means more world.

The hardest thing about making maple syrup is making the fire. The fire has to be very hot, hot enough to exceed the boiling point by 4 degrees Celsius, 7 degrees Fahrenheit. A densimeter is used to measure the density of the sap. Maple syrup is ready when 59 brix hot, 66 at room temperature, brix equivalent to the percentage of sugar.

The evaporator is 10 feet long and 3 feet wide. It has a large pan in back with folds on the bottom that sit in the fire, and two shallower front pans at the front, all stainless steel. The fire box is large—it fits 20–30 logs. You have to extend your arm deep into it when you throw the logs in, and you have to do it carefully, despite the need for speed: you can't hit the back pan. It's an art: easily 1000 degrees, the heat scorches your face as you reach in. Once the door is open, the movements have to be quick and precise. Ideally, the door stays open for only a minute or two. If the door stays open too long, you lose too much heat and it can take hours to build the fire back up. If the pieces of wood are too big, it burns too slowly. If you use too much softwood, it can be too quick to burn, and you have to open the doors too often. We learn to layer—larger pieces of hardwood followed by split hardwood and then crusts of softwood. This is why we split wood every morning. What you want is the heat of the embers, and so there has to be enough air to allow them to glow. But sometimes everything works against you: the atmospheric pressure changes, a storm coming in, and there is just no way to get the fire going with the kind of heat you need. Some days you leave having built fire after fire without having gotten to the magic brix. It's a test of patience.

The line leads in 100 Acres, and I realise, some weeks in, that I am not painting, but drawing. The difference between painting and drawing feels like a question of movement, and, in particular, of the movement of the line. Working as I do on a 5x5 foot frame, the movements of the line are body-sized. There is no escaping the breadth of gesture, and its modulating reach. There is a lot of open space. Open not in the sense of uncharted, unused, but in the sense of potentialised. I wonder at the way the drawing machine has generated this singular quality of dancing lines (Fig. 18.2).

The last two degrees are the hardest to reach. At that point, you are on the cusp of evaporation, the sap just on the edge of changing species. But with cool sap continually coming in from the outdoor vat, the front pans risk forever remaining just short of the density needed. Almost always, we opt to block the back pans, no longer allowing cool sap into the front pans. This means the front pans, no longer cooled by incoming sap, can more easily enter into the final evaporation phase. Although we have the tools for precise measure, we tend to rely most on the sensory changes in

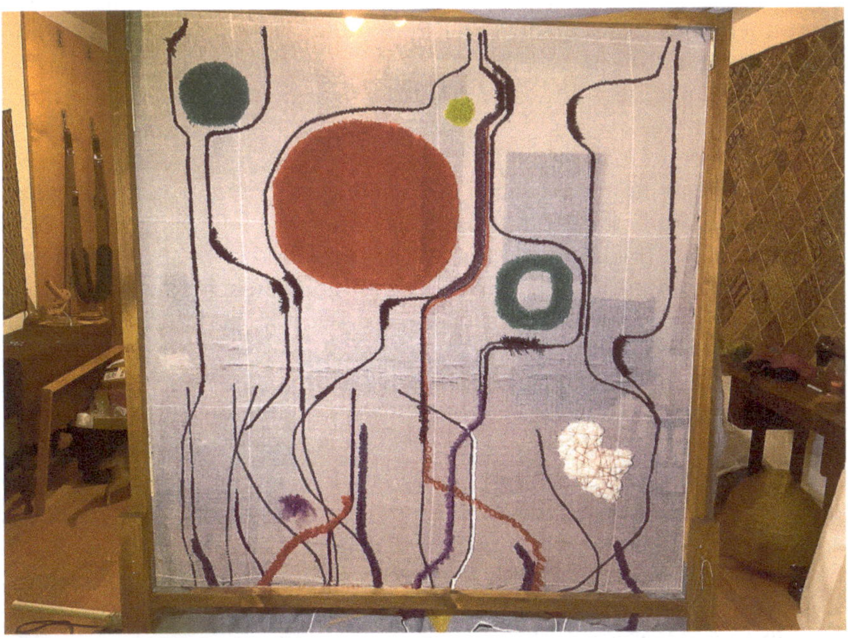

Fig. 18.2 100 Acres, Erin Manning

the surround: the smell of maple intensifies, the bubbles begin to get bigger, the sap begins to rise in the pans, risking boiling over, the sap becomes more golden. That's when we insert the large wooden spoon into the pan, checking to see how the sap drips from the spoon's wide surface. We are looking for what is called 'curtaining'—the effect of the sap dripping in a curtain-like formation from the edge of the spoon. In time, you can feel it when the threshold tips and sap turns to syrup. But that doesn't mean you always get it right. What we realise this year, having become better fire-makers, is that the sap stops curtaining when it goes above the required brix, when it is turning to taffy, and becoming sugar. This happens several times, and we see it in our bottles, which are crystallising. It's a tight threshold between sap, syrup and crystal, between under-firing and over-firing. The balance is hard to attain.

Often I have to ask Brian if a panel feels finished. It's an odd thing, making a continuous piece without a sense of what came before or what will come after. What I am looking for in each panel is the quality of commitment: does the surface come together in a way that feels as though it is both creating consistency and generating difference? Sometimes it's not quite there and I can feel it, but I find it takes an outside eye, a curious touch, to demonstrate where 100 Acres is calling for more tending, or restraint. Sometimes I sit with a panel for several days. Sometimes it takes an afternoon.

The process of making maple syrup is both incredibly quick and incredibly slow. Many hours are spent maintaining the conditions for evaporation, working to bring the heat up. It's not unusual for the temperature to stay around the boiling point for two or three hours while you stoke the fire, skim the impurities from the boiling sap, watch that it doesn't boil over, check the sap levels both inside and outside (particularly if the sap isn't flowing well and you have to be careful to have enough sap for the 40-1 evaporation). It is slow in the sense that it is repetitive, but everything can change so quickly that you have to always be poised for redirection.

The mapleing process, like all processes but emphatically so, does the directing. A shift in weather conditions can radically alter the fire, causing a surprise burst of heat which can deplete the store of sap to a level that is dangerously low. Or, a slightly wet piece of wood can dampen the fire, slowing things down. There are of course approximations born of repetition and we do move in the vista of expectation, yet it's not unusual to hear ourselves complain that 'yesterday the temperature was reached after

5 fires' while we stoke the seventh fire and continue to linger 2 degrees shy of our magic number. We spend a lot of time theorising about what is happening: maybe if we open the fire less often? Maybe if we make more smaller fires? Maybe if we use more softwood? Maybe if we shift the air vents? Close them? Open them?

Too much shifting of conditions works against us, but we don't always have the patience. Sometimes we just need to give up, head home, have a bath, eat dinner. Often that happens when we were sure, really, really certain, it would only take one more fire. But four fires later we are too impatient, too exhausted, to enjoy the process. So we leave. And start all over again the next day.

I do not, in the end, use all the remains. I weave as many of them as I can into the textile—begin every day with this practice—but ultimately there is just too much on the floor. I try making some cushions with the dregs but am unsatisfied. They are still in bags in the studio (Fig. 18.3).

The syrup is affected by every decision. If you leave after a long day of boiling without pulling a batch, the batch the next day will be darker. It will have boiled longer to get to the required sugar content. Whiteness has infiltrated syrup culture. Lighter syrup is sought after—it goes for a higher price. While there is a general consensus amongst syrup makers that the pale syrup is the least complex, the federation pays more for it, so everyone tries to make it. This involves cooking fast, in very clean pans. Less residue, less time, less complexity. Nothing bears this out. In fact, with blind taste tests, it has become apparent that colour is not even a defining factor when it comes to the taste of maple syrup. Some light batches can have the depth of dark ones and some dark ones can lack the complexity of light ones. The conditions are complex and cannot be simply reduced to transparency. No surprise there.

One of the most time-consuming things about cutting down trees is figuring out how to deal with the residue. At least half the tree needs to be disposed of—all those branches that aren't worth the time and energy to be cut for firewood or stripped for building materials. We build branch cemeteries. We make land configurations with them, filling in areas that have been cut into. But still there is always more than we know what to do with.

When trees fall in the maple forest, we don't take the wood out. We don't have a road and don't want to build one. And so we "waste" it. Habitat for insect and reptiles and amphibians, for mycelium and vegetation keeps the old growth forest alive.

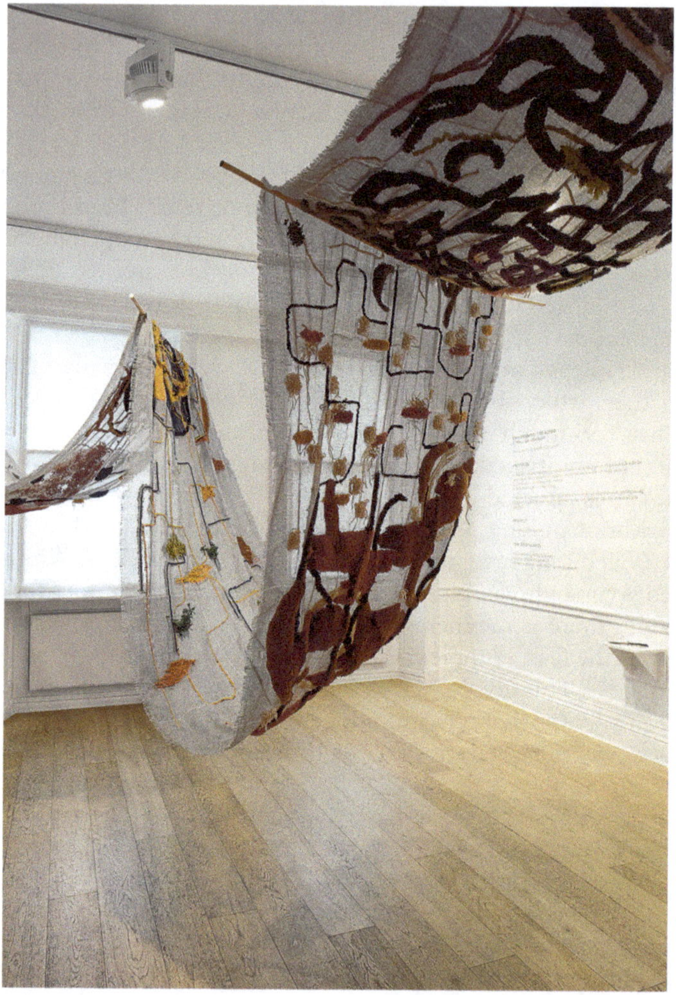

Fig. 18.3 100 Acres, Erin Manning at Richard Saltoun Gallery, London, 2024

The forest economy is never backgrounded in the 3Ecologies Project. We know the cost of burning wood, of cutting wood, of culling forests, of mapleing, of living off-grid. And we also know, from living in the city, the cost of pretending that electricity is cheap, abstracted as it is in Quebec from the fact of the hydro dams that participate in the ongoing decimation of indigenous lands. There is no dogmatic way, no binary path, to approach the environmental ecology. It needs to be engaged transversally, across the social and the conceptual. Most of us use more energy than we need because we have no idea of the cost of our need, or even what our need really is. With our hundred acres we are learning to gauge the cost, and the need.

The aesthetic yield of artfulness is not reducible to the art object. 100 Acres is not a representation of the forest. The aim is not to draw a comparison. The hope is to feel into the artfulness of what exceeds the gauge, of what meets the need in a shared ethos without return.

The 3Ecologies Project lives in the transversality of the uneasy metastability of three ecologies in co-composition. Often, we struggle to move beyond the very pragmatic necessities of its daily upkeep. We bring books to read at the sugar shack but never open them. The pull of the process is strong—even when there are moments of lull, they don't bring with them the quality of attention that would allow a book to be opened. Books are their own worlds, and the maple shack is too present in its imminent demands to give in to concurrent worlds. We may sit down, for a minute, but soon enough we are standing again, cloth in hand, wiping the side of the evaporator as we filter the back pan of impurities, checking the temperature and the boil on our way by.

Artfulness has a technicity to it. It outqualifies the sum of its parts, quality + quality.

We have found that boiling the sap just under the 59 brix produces a syrup with a more complex aftertaste. Going over 60 brix produces that delicious viscosity Quebecers love but sometimes the maple taste is too dominant, masking the earth-tones and the complex echoes of nut and vanilla. For locals, the viscosity is everything. There isn't much room in Quebec's syrup mafia to range, but we do it, appealing to the culture of tasting notes that accompanies practices such as wine-making or olive oil production.

Artfulness coincides—in the best conditions—with artmaking, but can't be reduced to it. Artfulness exceeds the work-as-object. It's that quality that tugs at the work's edges to move it in a direction it could not have anticipated, need meets excess, subtracted.

Walking the forest is to move in the vastness of intricate detail. If you look at your feet, you are soon stilled by all that grows there, and make very little headway. If you look ahead, pushing through the maze of brush, in the understory, you can lose sight of the underground rhizomes and root systems that travel the forest towards its vertical tendencies. In counterpoint, mapleing leads the way towards movements that forest.

In the gallery, 100 Acres is not accompanied by an artist statement. I have no statement to make about it: once hung on the trapezes, it will resituate itself in a new milieu, and I will have to become acquainted with it anew. In that space, against the temptations of distantism, the work will crowd you out if you attempt to be at arm's length. But if you approach it, if you enter into its folds, it will embrace you. You, I, will never be able to see the whole vista of it. It's just too big. And so you will feel as though the detail of the handiwork is lost, perhaps. But forests are like that—too vast to approach at scale. Only clearcutting does this, decimating the vastness to create overview (Fig. 18.4).

The mapleing has now ended, and it's time for the untapping. There is a small window for this if you want to avoid the blackflies. And believe me, you do.

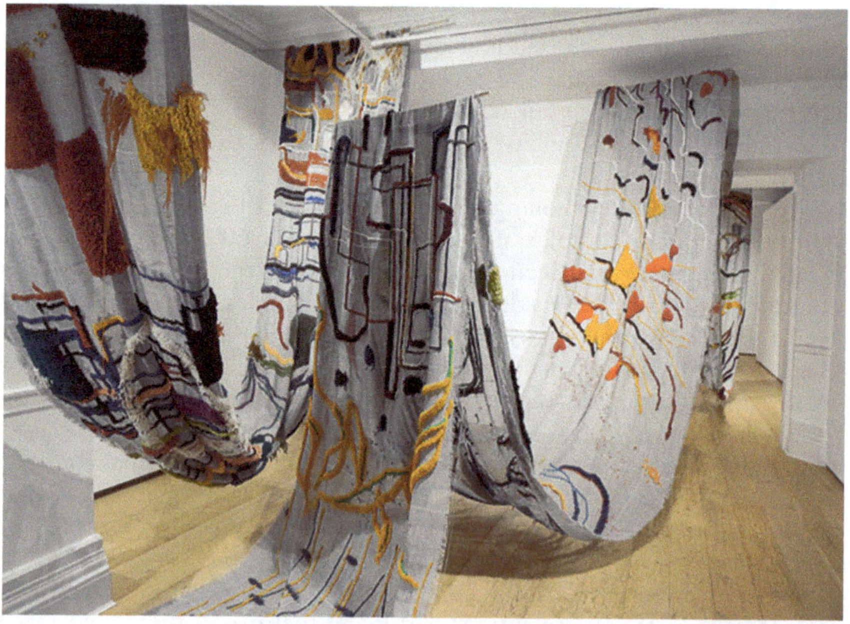

Fig. 18.4 100 Acres, Erin Manning at Richard Saltoun Gallery, London, 2024

Untapping involves moving from tree to tree, taking out each tap, and putting a small amount of alcohol in each tube, then closing it until the next season. Over the winter, the alcohol vapour will clean and disinfect the lines, and when the trees are tapped again the following spring, 2 inches over and 10 inches up or down, so as not to harm the tree, the first sap will flow into the ground, getting the alcohol out of the maze. And then you will close the system, and begin again: the sap will flow down the mountain into the vat, into the evaporator, you will make the fires, check the temperature, curtain, and pull the batch, each step of the process as if happening for the first time.

PROTOCOL

1. Negotiate a cut with the artist for extraction (4 ft. x 5 ft., £2000)
2. Coordinate a minor repair of the incision
3. Resituate the piece in a new milieu

All funds will go towards the purchase of additional land for the 3Ecologies Project (3ecologies.org). Purchased land will be taken out of the property market and placed in collective stewardship in perpetuity.

PROJECT

Give the land back to itself.

THE 3 ECOLOGIES

Of subjectivity (in its nascent state)
Of the social (in a mutant state)
Of the environment (at the point it can be reinvented)

Competing Interests The author has no conflicts of interest to declare that are relevant to the content of this chapter.

Reference

Manning, E. (2023). *Out of the clear. Minor Compositions.*

CHAPTER 19

Aesthoecology and the Earth Crisis: At the Intersection of Aesthetics, Ecology and Ethics

Chris Turner

Introduction

Engagement with such critical events as anthropogenic global warming that accentuates already existing acute social justice issues requires more than insertion into the curriculum in narrowly defined contexts. Opportunities to engage fully with critical global issues in the Anthropocene are essential. Personal and societal capabilities need to be developed to adapt to an unknown, insecure future with courage and optimism.

My research and teaching interests centre on education as aesthetic and ecological practice that values the affective, connected and ethical dimensions of the learning process, emphasising creative and radical approaches to problem-solving. This has led to advocacy for transdisciplinarity in the learning process. Although my current research is theoretical and philosophical, it is informed by, and draws upon, many years of practical

C. Turner (✉)
University of Exeter, Exeter, UK
e-mail: c.turner5@exeter.ac.uk

M. Matthews (ed.), *Ecologies in Practice and Learning: Arts Interventions in the Earth Crisis*, Palgrave Studies in Educational Futures, https://doi.org/10.1007/978-3-031-82338-1_19

experience in education. I propose that aesthoecology (Turner, 2019) as an onto-epistemology (Barad, 2007; Geerts & Carstens, 2019), a way of being and knowing, is an appropriate approach to futures-thinking (Canina et al., 2022; Facer, 2011) and understanding more fully, not just education, but global issues such as the Earth Crisis. It is an effective way of seamlessly blending education and global issues to mutual benefit. Adopting aesthoecology (Turner, 2019, 2024; Turner & Hall, 2021), a symbiotic relationship between aesthetics and ecology, is fundamental to critical educational engagement with the complexity of the Earth Crisis (Turner & Hall, 2023). Aesthoecology has much in common with Lovelock's conception of Gaia (Lovelock, 2000), a complex system of self-regulation maintaining life on Earth, combined with Naess's notion of deep ecology (Naess, 2008). I will elaborate on these concepts later.

An appreciation of aesthetics does not stand in isolation. Whether defined as perception, the arts, sensitivity, affectivity, empathy, emotion, wonder or beauty, or a complex combination of them all, aesthetics is implicated in a fascination and concern for the organic and dynamic nature of the planet and in eliciting a sense of connection with the material world.

Jane Bennett (2010), Graham Harman (2018) and Timothy Morton (2009, 2013) extend this connectivity through exploring a flat ontology, the agency of entanglements and the power of matter. 'Thing-power' (Bennett, 2010, p. 20) does not stress individuality but a 'congregational understanding of agency' (ibid.): an actant does not act on its own. It depends upon interaction and collaboration to provide a vital impetus. Other-than-human connections are illustrated through the writing of Donna Haraway (2016), Karen Barad (2007) and Rosi Braidotti (2013, 2019) on posthumanism. Braidotti, for example, works 'towards elaborating alternative ways of conceptualising the human subject' (2013, p. 37), critically challenging the binary of the human and the non-human (Braidotti, 2019).

Aesthoecology and Entanglement

The shape, form and composition of objects determine the way in which they are perceived and how they collaborate, acting less as social constructs and more as vital materialities, drawing their relationship with humans into the sphere of ethics and politics. Arguably, 'art *is* a material practice and … materiality of matter lies at the heart of creative practice.' (Barrett & Bolt, 2013, p. 5). This type of form-matter synthesis, however, implies that matter is, as Barrett and Bolt put it, dumb, mute or irrational and only comes to life in the hands of the artist in creating form. Matter becomes 'the substrate and field for the artist's formative action' as Heidegger (1977, p. 52) identifies it (Barrett & Bolt, 2013, p. 5).

If art, therefore, has a particular relation to the matter of things, then aesthetics, in all its forms, is a way of perceiving the world that comes alive through continuous reflection on conceptions of the world and our place in it. Interrogation and association are processes that compare the external environment as received through the senses with that which has already been processed and retained as neural memories. These processes lead to a conscious state of reality (Morais, 2019; Zhu, 2022). This is inherent in my proposition that the connection between the aesthetic (our perceptual and reflective) and ecological (our environmental) dimensions of learning and knowing is so strong that it can be considered to be continuously holistic—an aesthoecology.

Connectivity in this context is envisaged as complex, rather than linear, made up of assemblages (DeLanda, 2016), or what Timothy Morton (2013) would call a mesh, in which each point is both the centre and the edge of a system, so that centre and edge are entangled. Although Morton primarily problematises this theory from the standpoint of ecological entanglement, I consider this entanglement to possess aesthetic dimensions that arise from the idea that the components of a mesh are not considered as independent units but as phenomena defined by Barad (2007, p. 33) as 'the *ontological* inseparability of agentially intra-acting components'. Barad distinguishes between 'intra-action' and 'inter-action': in the former, 'distinct agencies do not precede, but rather emerge through, their intra-action' (ibid.).

Symbiotically binding aesthetics with ecology in this mesh brings about a reconceptualisation of other related philosophical concepts, not least that of ethics. As Harris (2016) suggests, 'Climate change is a profound matter of injustice imposed on people who are already poorly off and who usually have no say in the matter' (p. 37). The ethical context of this chapter is the Earth Crisis. However, a care for the Earth and the life that it supports incorporates far more than anthropogenic climate change. Intersectionality (Al-Fahim et al., 2019) analyses interlocking systems of power affecting those most marginalised in society and focuses on issues of climate justice and the ways in which we ethically relate to our respective environments (Kaijser & Kronsell, 2014). A term first used in 1989, intersectionality (Crenshaw, 1991) highlighted injustices experienced by women of colour and the complexity of intersecting characteristics, such as race, class and gender, that multiply and complicate issues of discrimination. Climate change is another multiplying factor because it will, at least initially, have greatest effect on those already disadvantaged, marginalised and discriminated against.

The ways in which we refer to climate change and the rapid decline of biodiversity are important. I question, for example, normative and historic terminology such as *nature* (Ducarme & Couvet, 2020). The careless way in which this term is used leads us into false conceptions of our status on the planet. According to Morton (2009, p. 1), consideration of a notion of *nature* will have 'to wither away in an "ecological" state of human society'. He suggests that the idea of *nature*, a nineteenth-century and romantic ideal, should give way to 'properly ecological forms of culture, philosophy, politics and art' (ibid.), and is another form of intersectionality. Morton proposes that ecological criticism must remove the division between nature and culture and challenge the idea that nature exists as something that supports and nourishes civilisation, but outside of society. As Berleant notes (1992, p. 15), 'the concept of the environment as outside, external to the human organism, is a comforting notion now utterly discarded both by ecological studies and post-Cartesian philosophy'.

The dualism of *man* and *nature* has been a distraction from an understanding of inter-dependence and inter-connection between all entities existing in and on planet Earth. This has led to an anthropocentric, isolationist view of the human condition that has resulted in a distortion of our understandings of ecology and led to exploitation, inequality and imbalance that can only lead to further degradation of living systems. There must be recognition of the way in which posthumanism, new materialism, ecology, ethics, politics *and* education intersect to create new ways of framing contemporary global issues.

This also applies to functions of art. Rancière (2004) refers to aesthetics not as a discipline but as a way of thinking about the arts. In common with politics, the aesthetic represents a pivotal balance between forms of action, production, perception and thought. Politics and art share the processes of challenging the status quo and re-ordering power relationships. This is why I propose that a consideration of aesthetics is fundamental to radical educational reform and essential for society to face the existential challenges that climate change brings. Aesthetics is central to transdisciplinary thought (Turner & Hall, 2021).

Aesthoecology, through an organic lens, looks beyond the human. It examines the nature of things and objects and the vibrant relationships existing between the human and the other-than-human, by intimately associating aesthetics and ecology. This may be seen to operate as a complex web of networks or assemblages that constructs an ontology of what is happening around us, with us and through us, such that:

> Our way of being in the world is a continuous transformative state—we redefine ourselves continuously, in relation to our surroundings, which we in turn create, shape and alter through transformative experience. (Mitrache, 2013, p. 545)

Education is an important dimension in understanding and challenging current ways of relating to the environment. In examining the role of aesthoecology as a way of knowing and being in the world, there is hope that actions may be taken to avert future degradation and damage (Fig. 19.1).

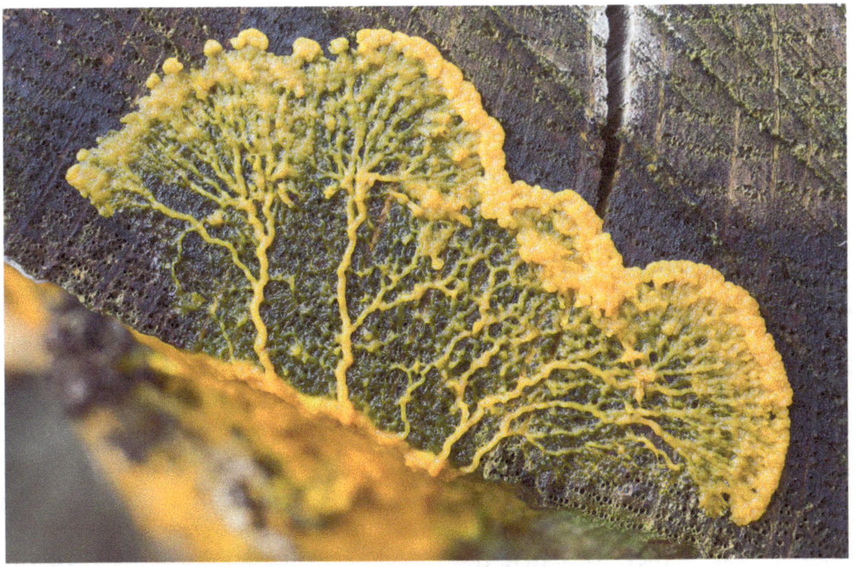

Fig. 19.1 Slime moulds: representing assemblages of aesthoecological networks that transform their environments

The Earth Crisis as an Entanglement of Wicked Problems

A transdisciplinary pedagogy may also be pivotal to the historic and contemporary issues that have beset the planet. They are of such a magnitude that they have become known as *wicked problems* (Rittel & Webber, 1973). Wicked problems are not individual, one-off issues; they possess a complexity that involves multiple possible causes, outcomes and internal dynamics that are far from linear (Chappell et al., 2024; Osberg & Biesta, 2008). The climate crisis is extremely complex (Sun & Yang, 2016), and difficult to find meaningful responses to. There are no individual problems that can be easily isolated and solved. Aesthetics provides a different, multi-dimensional focus that can stimulate creative solutions.

Ecological problems, particularly those related to climate, are seen as particularly wicked because each is seen as uniquely complex and located 'at the boundaries of natural and social systems' (Van Bueren et al., 2003, p. 193). Value judgements are often inherent in wicked problems (McBeth & Shanahan, 2004), thereby making them deeply ethical issues. For example, whether to impose lockdowns on all during Covid-19, or whether to protect the most vulnerable while allowing everyday life to continue for the majority was a key dilemma. Whichever path taken would have irreversible and unforeseen consequences. Bauman (2017) poses the question: 'What might a planetary ethic – beyond identity politics, national borders, religious borders, and species borders – look like?' (ibid., p. 8). Aesthoecology provides a prism through which we might think through planetary issues such as climate change and associated educational praxis.

If the lens through which these wicked problems are perceived assumes that the anthropocentric focus is a cause of most wicked problems, then a shift from an ego-centric perspective to an eco-centric perspective would give rise to some solutions—a different way of being and knowing. Change in relationality between the human and the more-than-human, recognising a more eco-centric perspective and flat hierarchy, might offer that fissure in normative thinking, those *creative ruptions* (Chappell et al., 2024), that allow us, the human, to value and respect these juxtapositions differently.

Wicked problems are similar to Morton's (2013) *hyperobjects*—entities of such vast dimensions that they challenge all traditional ideas of the nature of things and the relationship between them. An obvious Earth Crisis issue is global warming, and Connolly (2013, p. 410) adds other contemporary examples such as cross-regional violence, rapid migration of displaced peoples and complex loops between bees, viruses and pesticides.

The impact of hyperobjects, or wicked problems, is becoming apparent in diverse ways and cannot be combatted by some delusion that our subconscious awareness will not be lived out. The delusional aspect is that we can, on a day-to-day basis, push this awareness away because it is concerned with things of such vast dimensions that we do not know how to deal with them. The delusion is that fundamentally we know these complex problems are there, and they rise to the surface when we are reminded of them, but we must get on with our daily lives, despite the recognition that we continuously contribute to them:

> I do not access hyperobjects across a distance, through some transparent medium. Hyperobjects are here, right here in my social and experiential space. Like faces pressed against a window, they leer at me menacingly: their very nearness is what menaces. (Morton, 2013, p. 27)

I have a particular interest in anthropogenic climate change because I believe that these issues in the next decade will drive and exacerbate many other wicked problems. I contend that the relationship between aesthetics, ecology and ethics provides a framework for us to gain some understanding of these complex issues and, thus, to come terms with them. They are educational imperatives that demand radical and creative action.

Pedagogical possibility thinking (Cremin et al., 2006; Snaza & Weaver, 2015) arises through aesthoecological imagining. Climate issues are, at heart, issues of social and ecological justice, and education must stimulate creative-futures-thinking that radically alters perceptions (Jenkins et al., 2021) so that the preparedness of young people changes from a position of eco-futures-anxiety to one of eco-futures-action (Turner, 2024). I propose that if we were to subscribe to a more posthumanist and ecophilosophical understanding of educational relationality educators might decide on very different transdisciplinary pedagogical practices and research paradigms.

For example, Pacini-Ketchabaw et al. (2016) talk of numerous earthworms which cover the sidewalks in British Columbia after every period of rain. Groups of children and teachers observe these areas, pay attention to where they place their feet, heightening awareness of the importance of avoiding harming these creatures. The children's hands reach out to touch the earthworms. 'The worms wiggle; the hands gently hold. Every embodied encounter changes both' (ibid., p. 154). This is a reminder that:

> There are innumerable threads that knit our common worlds together, including these small chance encounters of children and worms on the slippery pathways of everyday life. (ibid.)

I was recently invited to take part in a group activity for teacher educators and researchers, at the Islington Ecology Centre, London, which is sited on a small nature reserve—a former railway station sidings that has been allowed to rewild. It is used extensively to provide children of Islington and beyond with an experience of natural habitats and biodiversity, to get close to a wide variety of organisms, to study them and to develop an imaginative, immersive and embodied relationship with them. Artwork, storytelling and relating, in an embodied sense, to the environment are activities to stimulate wonder, empathy and appreciation (Schinkel, 2022). It was inspirational.

These examples of pedagogy represent an onto-epistemological shift—different ways of being, knowing and focusing—rupturing education's fixation on linear measurement and directive systems to that which more fully values affective, connective and ethical relationship transformation.

Re-aligning Understandings of Relationality Through Posthumanism and New Materialism

Drawing on the notion of Morton's entangled web, it is not until we transition from the centring of the human, the anthropocentric position, to one that accepts greater equity with what Braidotti (2019) calls the other-than-human and Taylor and Hughes (2016, p. 13) describe as 'an ecology of human-nonhuman relations in which we (all) are embedded and tangled,' that we may start to understand how we got to where we are and find some pathway out.

Re-ordering the importance of relationships, calling on the notion of sympoiesis, meaning *making-with* or *becoming-with* (Witt & Clarke, 2024), is essential in steering us towards different educational imperatives and pedagogies. Witt and Clarke's reference to collective making emphasises relationality at the heart of educative practice and the importance of creating spaces to do so. They illustrate the importance of both aesthetics (being with and being sensitive to) and ecology (stressing connectivity) in a more-than-human world. Their pedagogical research creations illustrate the power of art as a collage of writing, photographs and patterns to illustrate theoretical concepts of sympoiesis.

This emphasis on re-ordering relationality and affectivity requires alternative philosophical and theoretical models, to adopt a new onto-epistemology to better understand and tackle the new order of challenges that the educator and others are facing. I propose that aesthoecology is suited to that purpose (Hall & Turner, 2021; Turner, 2019, 2024).

Aesthoecology, Gaia and Deep Ecology

Aesthoecology has much in common with Lovelock's conception of Gaia (Lovelock, 2000) and Naess's notion of deep ecology (Naess, 2008). Aesthoecology has been derived much more from an educational perspective and has spread outwards, whereas Gaia is a scientific concept which was, at a later stage and very importantly, disseminated through education, for example, at Schumacher College, Devon, and the Global Systems Institute at the University of Exeter. The incorporation of aesthetics, central to deep ecology (Naess, 2008), adds an important dimension that has both an epistemological and pedagogical significance, making it a powerful agent in determining how we act in the world (Chave, 2020).

Interpreting the aesthetic in the context of deep ecology encourages humans to live less violently in their relationships, for example, by stopping the use of toxic sprays and fossils fuels, reassessing our use of land, and bringing our activities into harmony through a more intensive understanding of what relationships within the natural world entail. Supporters of the 'shallow ecology movement', Naess claims, 'do not question deeply but focus on short term, narrow, human interests' (Naess, 2008, p. 26).

Art education, in the context of the environmental crisis, that fosters a critical and questioning approach to ecological issues, might be seen as 'deep ecology art'. That which depicts the world purely in romantic terms, or simply in representational, familiar ways, neglects the seriousness and cutting edge of activist art. The cutting edge, as interpreted by Rancière (2004), is a form of dissensus, or disagreement, in which 'Art can become life. Life can become art. And art and life can exchange their properties' (p. 119). This is a form of metapolitics 're-configuring art as a political issue or asserting itself as true politics' (ibid.), and, I suggest, is important as deep-ecology art. Its basic premise is that matters of art are matters of education and, in an aesthoecological sense, create, in practice, a new and ethical regime.

Deep ecology more fully focuses, in a sincere and empathic way, on an understanding of the intrinsic worth of being in relation with the world, deeply valuing the enmeshed qualities of those relationships and throwing off the cloak of selfishness and insularity. Naess stresses that the richness and diversity of life have intrinsic value and 'that, except to satisfy *vital* needs, humans do not have the right to reduce that diversity and richness' (Naess, 2008, p. 28). Further than that, there is a growing recognition that the diversity and richness of collective ecologies, human and more-than-human, is essential to the future of the human species, not just a bonus.

In accepting the creative potential of all matter, normative anthropocentric thinking can be disrupted, and transformational thinking adopted. By focusing on relationships with objects, hierarchical perceptions change. This acute awareness of organic and dynamic materialism fosters socially responsive and sustainable practices, which heighten new understandings of ethical relationships. This is a recognition that the material world is not a fixed entity but is continually in flux (Barad, 2007; Coole & Frost, 2010) and has the capacity for agency that extends beyond that of humans (Braidotti, 2013, DeLanda, 2016, Latour, 2005).

New materialists who ascribe agency to matter, such as Jane Bennett (2010), are attracted to ideas of vitalism that refute the sharp distinction, generally made by physicists and biologists, between inorganic matter and organic, living systems. Barrett and Bolt (2013), citing Poe (2011), engage with the question of whether new materialism can account for the aesthetic dimensions of objects. Poe sees the human as the active creator in the production of art: what Heidegger (1977, p. 52), when considering matter, calls 'the substrate and field for the artist's formative action'—the form-matter synthesis. For Heidegger (according to Barrett & Bolt, 2013) the relationship between artist or maker, and things or objects, is one of 'co-responsibility and indebtedness' (Barrett & Bolt, 2013, p. 6) and moves on from the form-matter synthesis to one in which 'art is a co-collaboration, not a form-matter synthesis and matter as much as the human has responsibility for the emergence of art' (ibid.). This supports Bennett's proposal (2010) that matter has agency. In relation to aesthoecology, the aesthetic experience is not just resident in the human subject, but 'situates the aesthetic as a relationship "between" the human and non-human, the material and the immaterial, the social and physical' (Barrett & Bolt, 2013, p. 6)—a symbiotic relationship between the aesthetic and the ecological. Therefore, aesthoecology challenges anthropocentric approaches to cultural practice and the dualist ontologies of mind-body, self-world and nature-culture (Fox & Alldred, 2017).

Conclusion

To manage education, rather than to truly understand it, seems to assume structures, timelines, constraints, measurements, hierarchy, control and oversimplification. That is understandable if what is required follows a reductive model of education. However, if an education system is to provide an organic, adaptable system for living with, and finding solutions to, the Earth Crisis that faces us, a much more imaginative, artistic, flexible way of approaching education is required, and required urgently.

We no longer have the luxury of implementing minor changes to education, or even yet another round of so-called curriculum reform that still focuses on a schooling system designed for a very different age. Although I have primarily focused on UK education in this chapter, the worrying events of which I write are of global importance. Acting, or activism, in the world is essential at this time of increasing numbers of Earth crises predominantly arising from damage inflicted on the planet by anthropocentric activity. Hope is important (Freire, 2021), and courage is an active progression from hope that entails confidence in our convictions, informed by scientific and secure evidence and a willingness to share those beliefs—in short, to step up and be counted at the earliest opportunity.

According to Garcia-Gibson (2023, p. 1), climate activism is 'a confrontational activity that contributes (or whose actors believe or claim that it contributes) to preventing (or to expressing opposition to) impersonally bad or unjust outcomes', many of which are the result of anthropogenic global warming. We require a new educational culture. This will require a transdisciplinary, lifelong pedagogy that actively seeks spaces and opportunities to enact aesthoecological change. Aesthoecology underpins an approach that encourages different relationships with the environment aesthetically, ecologically and ethically. This must be reflected continuously through emergent education systems in the wake of planetary and ecosystem decline.

Competing Interests (4) The author has no conflicts of interest to declare that are relevant to the content of this chapter.

References

Al-Fahim, H., Davis, A. M., & Ernst, R. (2019). Intersectionality: From theory to practice. *Annual Review of Law and Social Sciences, 15*(1), 247–265.

Barad, K. (2007). *Meeting the universe halfway*. Duke University Press.

Barrett, E., & Bolt, B. (Eds.). (2013). *Carnal knowledge: Towards a 'new materialism' through the arts*. I. B.

Bauman, W. A. (2017). The ethics of wicked problems: Entanglement, multiple causality and rainbow time. *Worldviews: Global Religions, Culture, and Ecology, 21*(1), 7–20.

Berleant, A. (1992). *The aesthetics of environment*. Temple University Press.

Braidotti, R. (2013). *The posthuman*. Reality Press.

Braidotti, R. (2019). A theoretical framework for the critical posthumanities. *Theory, Culture & Society, 36*(6), 31–61.

Bennett, J. (2010). *Vibrant matter – A political ecology of things*. Duke University Press.

Canina, M., Bruno, C., & Monestier, E. (2022). Futures thinking. In *The Palgrave encyclopedia of the possible*. Palgrave Macmillan.

Chappell, K., Turner, C., & Wren, H. (2024). *Creative ruptions for emergent educational futures*. Palgrave Macmillan.

Chave, S. (2020). *Sustainable and democratic education: Opening spaces for complexity, subjectivity and the future*. Routledge.

Connolly, W. (2013). *The fragility of things—Self organizing processes, neoliberal fantasies and democratic activism*. Duke University Press.

Coole, D., & Frost, S. (2010). *New materialisms: Ontology, agency and politics*. Duke University Press.

Cremin, T., Burnard, P., & Craft, A. (2006). Pedagogy and possibility thinking in the early years. *Thinking Skills and Creativity, 1*(2), 108–119.

Crenshaw, K. (1991). Mapping the margins: Intersectionality, identity politics, and violence against women of colour. *Stanford Law Review, 43*(6), 1241–1299.

DeLanda, M. (2016). *Assemblage theory*. Edinburgh University Press.

Ducarme, F., & Couvet, D. (2020). What does 'nature' mean? *Palgrave Commun, 6*(14). https://doi.org/10.1057/s41599-020-0390-y

Facer, K. (2011). *Learning futures: Education, technology and social change*. Routledge.

Freire, P. (2021). *Pedagogy of hope: Reliving pedagogy of the oppressed*. Bloomsbury.

Fox, N., & Alldred, P. (2017). *Sociology and the new materialism: Theory, research, action*. Sage Publishing.

Garcia-Gibson, F. (2023). The ethics of climate activism. *WIREs, 14*(4), e831.

Geerts, E., & Carstens, D. (2019). Ethico-onto-epistemology. *Philosophy Today, 63*(4), 915–925.

Hall, E., & Turner, C. (2021). Aesthoecology and its implications for art and design education: Examining the foundations. *International Journal of Art and Design Education, 40*(4), 761–772.

Haraway, D. (2016). *Staying with the trouble: Making kin in the Chthulucene (experimental futures)*. Duke University Press.

Harman, G. (2018). *Object-oriented ontology—A new theory of everything*. Pelican Books.

Harris, P. G. (2016). *Global ethics and climate change*. Edinburgh University Press.

Heidegger, M. (1977). *The question concerning technology and other essays*. Garland.

Jenkins, N., Ritchie, L., & Quinn, S. (2021). From reflection to diffraction: Exploring the use of vignettes within post-humanist and multi-species research. *Qualitative Research, 21*(6), 975–989.

Kaijser, A., & Kronsell, A. (2014). Climate change through the lens of intersectionality. *Environmental Politics, 23*(3), 417–433.

Latour, B. (2005). *Reassembling the social: An introduction to actor network theory*. Oxford University Press.

Lovelock, J. (2000). *Gaia – A new look at life on Earth*. Oxford University Press.

Naess, A. (2008). *Ecology of wisdom*. Penguin.

McBeth, M. K., & Shanahan, E. A. (2004). Public opinion for sale: The role of policy marketers in Greater Yellowstone policy conflict. *Policy Sciences, 37*, 319–338.

Mitrache, A. (2013). Spatial sensibility in architectural education. *Procedia—Social and Behavioural Sciences, 93*, 544–548.

Morais, I. (2019). *Aesthetic realism*. Palgrave Pivot.

Morton, T. (2009). *Ecology without nature*. Harvard University Press.

Morton, T. (2013). *Hyperobjects*. University of Minnesota Press.

Osberg, D., & Biesta, G. (2008). The emergent curriculum: Navigating a complex course between unguided learning and planned enculturation. *Journal of Curriculum Studies, 40*(3), 313–328.

Pacini-Ketchabaw, V., Taylor, A., & Blaise, M. (2016). Decentring the human in multispecies ethnographies. In C. A. Taylor & C. Hughes (Eds.), *Posthuman research practices in education* (pp. 149–167). Palgrave Macmillan.

Poe, A. (2011). Review essay: Things beyond objects. *Journal of French and Francophone Philosophy, 9*(1), 153–164.

Rancière, J. (2004). *The politics of aesthetics – The distribution of the sensible*. Continuum.

Rittel, H. W. J., & Webber, M. M. (1973). Dilemmas in a general theory of planning. *Policy Sciences, 4*(2), 155–169.

Schinkel, A. (2022). *Wonder and education: On the educational importance of contemplative wonder*. Bloomsbury.

Snaza, N., & Weaver, J. (2015). Introduction: Education and the posthuman turn. In N. Snaza & J. Weaver (Eds.), *Posthumanism and educational research* (pp. 1–16). Routledge.

Sun, J., & Yang, K. (2016). The wicked problem of climate change: A new approach based on social mess and fragmentation. *Sustainability, 8*(12), 1312.

Taylor, C. A., & Hughes, C. (Eds.). (2016). *Posthuman research practices in education*. Palgrave Macmillan.

Turner, C. (2019). *Education as aesthoecology*. Doctoral thesis, University of Exeter.

Turner, C. (2024). Exploring aesthoecology: Affective anticipation, liminality and emergence as features of alternative educational futures. In K. Chappell, C. Turner, & H. Wren (Eds.), *Creative ruptions for emergent educational futures* (pp. 51–70). Palgrave Macmillan.

Turner, C. & Hall, E. (2021). *Transformation through aesthoecology: Affectivity, connectivity, and the role of art in promoting transdisciplinarity*. Innovative Practice in Higher Education, *GLAD-HE Special Edition*. University of Staffordshire.

Turner, C., & Hall, E. (2023). Drawing and the global climate emergency. In *Art, sustainability and learning communities: Call to action* (pp. 141–161). Intellect Books.

Van Bueren, E. M., Klijn, E. H., & Koppenjan, J. F. M. (2003). Dealing with wicked problems in networks: Analyzing an environmental debate from a network perspective. *Journal of Public Administration Research and Theory, 13*(2), 193–212.

Witt, S., & Clarke, H. (2024). (Ma)kin(g) sympoietic more-than-human futures. In K. Chappell, C. Turner, & H. Wren (Eds.), *Creative ruptions for emergent educational futures* (pp. 123–153). Palgrave Macmillan.

Zhu, Z. (2022). *Chinese aesthetics in a global context*. Springer.

CHAPTER 20

Cosmopedagogies

Dennis Atkinson

Introduction

We tend to think that children or students engaged in study or learning practices in classrooms, studios or laboratories are doing so together in the same place at the same time. In contrast we might think of such spaces as a disjunctive synthesis of different place-times, thus indicating the divergence and difference, the different rhythms, of each student's (and a teacher's) place-time of practice. The tendency to think of such spaces in terms of a uniform space and time relates to our ways of coping with complexity, of how, for example in this case, to give order and regulation to our experiencing of classroom situations. Such ordering is constituted by a panoply of standardised templates in the form of pedagogical standards, curriculum content, learning objectives, assessment criteria and so on. In other words, such templates reduce the compositions, rhythms and individuations of different and divergent place-times of practice to a standardised format through which they are conceived. We require such conceptual and practical templates or devices to provide a sense of structure to pedagogic work but, and

D. Atkinson (✉)
University of London, London, UK
e-mail: d.atkinson@gold.ac.uk

M. Matthews (ed.), *Ecologies in Practice and Learning: Arts Interventions in the Earth Crisis*, Palgrave Studies in Educational Futures, https://doi.org/10.1007/978-3-031-82338-1_20

this is a key point of this chapter, if they become too rigid or prescriptive, they may advantage some whilst simultaneously disadvantaging others. If we recognise classrooms or other educational spaces as a disjunctive synthesis of different place-times of practice, and if we accept that generic or standardised criteria may recognise and value some modes of practice but occlude and marginalise others, then what kind of pedagogies might we consider more appropriate to the implied difference and divergence? My answer is to consider the notion of cosmopedagogies as a working notion to inform this pedagogical problematic of trying to respond effectively to the different ecologies of practice in which learners are engaged. I will tackle this problematic in the context of art education.

I remember vividly working in high school classrooms/studios teaching art, sometimes succeeding, sometimes failing; sometimes seeming to offer advice which worked, sometimes not; sometimes not knowing how to proceed with practices which appeared strange or mysterious. But were my actions or words at the time commensurate with a student's or child's pathway of practice? Or did my actions or my assumptions about practice, my guidelines, come between myself and the reality of the situation in which I found myself having to respond? Sometimes this may not matter that much but sometimes it does. This chapter will articulate a number of scenes that support an advocacy for the notion of cosmopedagogies as a working concept for creating pedagogical practices that are able to respond effectively to the divergence and difference of students' learning pathways and what they may become capable of. Such responses then imply that pedagogies themselves diverge and transform as they encounter and respond to practices that may initially feel ineffable but which require the development of an art and manner of pedagogic practice that enable them to flourish.

Scene 1: Art Practice: Instauring Cosmicities, Work-to-Be-Made

Isabelle Stengers (2006, p. 15) writes

> When a teacher feels that what she is doing is important, that it is not only a transmission of useful knowledge, [...] she indeed participates in what may be called a cosmic adventure, because the manner the children will experience new possibilities, feelings and ideas, or stubbornly keep to their abstractions, to their judgement about what matters and what does not, is indeed a cosmic stake.

If we accept that difference is what we have in common, then the pedagogical task of working with and responding to difference can be viewed as a cosmic adventure. In his book, *The Different Modes of Existence,* Etienne Souriau (2015) was concerned with a plurimodality of manners of being, or a plurimodality of *arts of existing*. He was particularly concerned with those more precarious, less confident or marginalised existences and their unrealised possibilities or potentials. David Lapoujade (2021, p. 4) acknowledges that for Souriau, 'we can only reach being through the manners in which it is given,' or through *its particular art of being*. Souriau gives an interesting little anecdote to express this concern in his text: *To Have a Soul: Essays on Virtual Existences* (1938), which has deep relevance for the ecologies of pedagogies of taking care. Lapoujade (2021, p. 25) provides a translation of this anecdote on which I draw.

Souriau describes a young child who carefully and at great length arranges a variety of objects on his mother's table, making sure that the arrangement is pleasing so as to give his mother 'great delight'. When mother arrives, she takes one of the objects that she needs and moves others to their usual place, thereby undoing the child's arrangement. When the child's sobs give way to desperate explanations which reveal to the mother her error, she exclaims with deep regret, 'Oh! My poor little angel, I hadn't seen that there was something there!' But, what was it that she did not see? What did the mother fail to see? She saw the objects but what she missed was the mode of existence that they are given by the child's point of view. She failed to see the 'architectonic that they sketch out in the child's eyes' (Lapoujade, 2021, p. 25). What the mother failed to see in the initial arrangement of objects was the child's point of view: a virtuality that she failed to perceive.

This deep concern for the plurimodality of 'manners' or 'arts of being' and the notion of 'points of view' (perhaps an early precursor to the 'perspectivism' of Viveiros de Castro (1998), I find useful for reflecting upon pedagogical practice and the different modes of existence involved. Souriau proposes an 'existential pluralism' such that all beings exist and feel in their particular manner and that beings have multiple modes of existence, as a body, a psyche, a memory, a sensation and so on, so that in this sense beings are viewed as plurimodal. Souriau's different modes of existence posit a pluriverse, a plurality of arts of existing. This does not imply a world of individual, discrete or independent beings but rather a world of modes of existence-in-relation, which has implications for practices of convivial belonging, whose plural components are constantly interacting and evolving in parallel with their environments. Souriau's interest in those precarious modes of existence, modes that are not recognised or are marginalised (Lapoujade, 2021, pp. 4–9), suggests a strong ethical and political dimension, a concern for rights. We might indeed, reflect upon the notion of a pedagogy of rights, a pedagogy that can extend to wider issues such as the Earth Crisis and a call for a people to come.

Of chief concern for Souriau is the process of *the-work-to-be-made*, which he chooses to exemplify in what some might think a rather outdated practice, a lump of clay being worked on by a sculptor. By illustrating the dynamic of the work-to-be-made through the process of sculpting clay, Souriau is not describing a *project*, a process informed by a pre-constituted plan, he is proposing an unplanned *journey*. It's not a case of a work to be made according to a plan or model. It's not a realisation of potential through the creativity of an inspired artist. It is more of a co-respondence, a co-responding of the to-be-made and the made. Nothing (artist or work) is given in advance; everything emerges along the journey.

A crucial part to be played along this journey is the role of what Souriau calls 'the virtuals', a term taken up much later by Deleuze and Deleuze and Guattari. We can perhaps liken the virtuals mode of existence to a cloud of potentials, 'a number of sketches or starts' (Souriau, 2015a, p. 156). Their nature is to be incomplete, awaiting an art that can make them exist otherwise in a different mode. We might consider the practice of making a painting. Is it not the case that such practice relies upon such clouds of potentials, that they are immanent to such practice, that they nourish it by opening up possibilities for practice? By acknowledging the mode of existence of the virtuals, all reality becomes incomplete in that it is always in between modes, it is not 'in' beings but 'between' them.

Souriau is not describing a process initiated by an artist, this journey has no pilot or conductor. It is not a case of a project emerging through trial and error but something much more vertiginous. At each moment the process of the work-to-be-made is precarious and so is the being of the artist. The different facets of both the work-to-be-made and the work, more or less complete, constitute the problematic of what he calls ***instauration***. It is as relevant to all practices, including pedagogic practice, as it is to the practice of the artist.

Instauration, from the Latin *instauratio* or *instaurare*, suggests recommencement, restoration, resumption of something unable to be completed or brought to fruition the first time. In other words, it denotes a complex journey involving different modes and arts of existence (actual, virtual, material, sensible, imaginary) involving artist and work and the cloud of possibilities that emerge and disappear continually. Souriau prefers the term instauration to creation in that creation tends to presume a prior creator, an artist, whilst instauration involves a complex dynamism of forces in which both artist and work become simultaneously. There is no prior artist or work, they both emerge or become simultaneously through the instauration of practice. The process of making is constantly under threat of failure as each move is a process of experimentation and simultaneously the work questions the maker; each phase of the practice can be seen as a proposition of existence (2015b, p. 229). Souriau (2015b, p. 231) writes:

> To instaur is to follow a path. We determine the being to come in exploring its path. In blooming, the being demands its own existence. In all of this, the agent must yield before the work's own will, must work out what it is it wills, and must renounce himself [herself/themself] for the sake of this autonomous being.

The work-to-be-made installs a constant questioning, it constitutes a problematic. It does not provide the artist with directions (nor does the artist have a clear plan). Souriau (2015b, p. 232) puts it this way:

> For let us not forget that the work's effect upon the [artist] never takes the appearance of a revelation. The work-to-be-made never says to us: "Here is what I am, here is what I should be, a model you have only to copy." Rather, it is a mute dialogue in which the work seems enigmatically, almost ironically to say: "And what are you going to do now? With what actions are you going to promote or deteriorate me?"

We might say that instaurative acts produce the possibility for a new cosmicity of practice. I am using the term cosmicity to refer to the organisation or the architectonic of a world. Such organisation and structure are not constituted by an individual but emerge from relational assemblages of beings in their particular milieus. In philosophy, the conceptual frameworks developed by philosophers such as Plato, Descartes, Kant, Deleuze, Arendt, Butler, Stengers, Haraway, within their respective milieus, instaur different cosmicities. Similarly, we can say that artists in their particular milieus instaur different cosmicities through their gradual unfolding of practice. The cosmicities of Jackson Pollock, Barbara Hepworth, Yayoi Kusama, Frank Bowling or Oscar Murillo are not the same. This process of instauration can also be conceived in children's drawing and other students' art practices, in the gradual establishment of new organisations and relations of practice that open up new worlds of practice, new cosmicities.

To repeat, there is no plan involved, no template to follow, but an ongoing journey that engages a co-responding of artist and work on different levels of existence within particular milieus. A co-respondence is that in which the work and the artist each make the other exist, they institute or instaur each other, and with every new episode of practice emerge new co-respondences. This has important implications for pedagogic practice both in relation to children's and older students' processes of instauration, (of the becoming of practice and the becoming of self, which are indissoluble) but also to those of a teacher, as these processes lead to the instauration of new pedagogic cosmicities. To undermine, marginalise or deprive a mode of existence is to deny its nature and act as though it has no reality, no legitimacy. This resonates with the current scandalous lack of SEN (Special Educational Needs) provision in England. Here the notions of ontological pluralism and *perspectivism,* discussed earlier, become important. In this light the anti-ableist programme of NSEAD and the work of Claire Penketh (2023) and the team should also be mentioned here in relation to the struggle for the democratising of neuro-diverse cosmicities as an insurrectional mode of pedagogic demands, that is to say the right to have rights and the power this then bestows.

Scene 2: Earth and World

In their final book together, *What is Philosophy?* (1994), Gilles Deleuze and Felix Guattari include a chapter entitled Geophilosphy (pp. 85–113). They begin with the following statement.

> Subject and object give a poor approximation of thought. Thinking is neither a line drawn between subject and object, nor a revolving of one around the other. Rather, thinking takes place in the relationship of territory and the earth. (1994, p. 85)

In this scene I explore the relation between territory (or world) and earth in relation to art practice and pedagogical practice. Deleuze and Guattari's notion of geophilosophy, where 'geo' refers to 'earth', abandons the phenomenological relation between a subject and world in which the former occupies a transcendent position in relation to the latter (Roberts et al., 2022, p. 137). In contrast Deleuze and Guattari, through the notion of geophilosophy, propose a non-phenomenological approach in which the phenomenological relation between subject and world is displaced by the notion of earth as a deterritorialising force that 'goes beyond any territory' (1994, p. 85). Deleuze and Guattari thus propose (p. 86) an ongoing process of deterritorialisation from world to earth and a reterritorialisation from earth to world. World is already constituted, in human terms, by established practices and values that pertain to different spheres of life such as politics, education, economics, ethics, science, aesthetics and so on. Earth on the other hand refers to an immanent plane and flux of forces that both precede and exceed the subject and world. In simple terms world constitutes a territorialised entity whilst earth is a deterritorialising force that may provoke new reterritorialisations. In relation to the ongoing nature of subjectivity, we can conceive this process as always under construction, or in Souriau's terminology, as always a work-to-be-made. This constant process of renewal is pictured by Deleuze and Guattari as the creation of a people to come or a becoming-people (1994, pp. 109–110).

The idea of a people to come has direct implications today for many spheres of human activity, including education. If we reduce education to 'world', that is to say, to established modes of practice and meaning, this would seem to dismiss encounters with forces of the planet Earth that lie beyond such parameters. And yet it is through encounters with the forces of earth, with that which lies beyond our established frameworks of understanding, that we are able to come to see their limitations. Taking this further, such revelations may also precipitate the affect of shame, an affective assemblage that may lead us to see our limitations, our biases, our prejudices, our conformist attitudes, our confident demeanours, in order change and transform practice.

Deleuze and Guattari identify the deterritorialising forces of earth upon our established horizons of practice with the practices of art, more specifically with the shocks of sensation that we can experience when encountering art or equally through the affects and struggles of making art. In such encounters the relation between artist and work is to be viewed as a relation between world and earth and not in terms of subject and object. In the encounter between world and earth, it is not a case of a subject (artist) perceiving or experiencing a world from his or her point of view, but rather the chaotic forces of intensities and sensations beyond everyday experiences that lead to the ontogenesis of a subject and a new world. As Roberts et al. (2022, p. 138) argue, 'It is through our encounters with art, and the mutant forms of subjectivity it produces that we can become capable of encountering the earth as that which precedes and exceeds the world.' Experience then can be thought more extensively beyond the notion of a subjective position constituting the horizon through which the world is given, because this excludes those forces that lie beyond such horizons, forces that arise from the chaosmos of earth. The task for Deleuze and Guattari is to try to think beyond the parameters of world and the limits these place upon subjectivity in order to engage with the forces of the earth and their potential modes of expressivity which, in turn, lead to the creation of a people yet to come and a reterritorialised world. In the context of educational practices in order to try to respond effectively to the difference and divergence of practices of learning, we might adopt the notion of *geopedagogies*. This would entail a similar endeavour to move from the limitations of established pedagogising forces that presuppose established worlds. Here 'geo' does not refer to such established worlds of pedagogy but to those forces and sensibilities beyond that deterritorialise such worlds and to learners and teachers yet-to-emerge.

The radical force of art practice is one that challenges and disrupts established practices and opens up possibilities for new practices to emerge. The pedagogical challenge in the context of art education is that whilst acknowledging and valuing established modes of practice, it has to acknowledge this radical force and its emerging possibilities and potentials. This does not just relate to the work of artists but also to the emerging practice of children and students, who in their learning encounters can open up new modes of thought and practice, new cosmicities. We might then view art practice as beckoning and celebrating what Isabelle Stengers calls *cosmos*, the focus of the next scene.

Scene 3: Cosmopedagogies

In her essay, *The Cosmopolitical Proposal*, Isabelle Stengers (2005a) writes that the term *cosmos*, 'must be distinguished from any particular cosmos, or world,' and that 'in the term cosmopolitical, cosmos refers to the unknown constituted by these multiple, divergent worlds and to the articulations of which they could eventually be capable' (p. 995). It is this emphasis upon *cosmos*, not to denote an established world, or worlds, but to conceive it as a speculative adventure referring to the unknown, divergence, multiplicity and capability that I find absorbing in relation to pedagogic practice and its insistent problematics. We might say that the 'unknown' of cosmos is closely related to the notion of earth proposed by Deleuze and Guattari and their notion of a people to come. Such problematics often arise when, in the presence of diverse worlds, some of which may be resistant or impervious to our pedagogical initiatives, we have obligations to respond effectively and empathetically to such diversity and 'the articulations of which it may be capable.' It is Stengers' idea of cosmos that underpins the notion of cosmopedagogies. Pedagogies whose obligation and concern are towards the divergent worlds of children/students and towards the practices they may be capable of.

The importance of Stengers' proposal for pedagogic practice, is that it prioritises the concrete situations of practice, their local sense of importance and the coherent narratives that ensue through which practice is constructed and conceived. It stresses the *importance* of the practitioner and the diverse situations in which they function. By implication it also advocates a need to resist the imposition of what we might call the opinions and theories of 'attendant experts' who may not be consonant with the concrete particularities of a practice. Stengers (2005a, p. 994) writes:

> the "cosmopolitical" proposal, as I intend to characterize it, is not designed for generalists; it has meaning *only in concrete situations where practitioners operate*...practitioners [such as teachers] who [can or] have learned to shrug their shoulders at the claims of generalising theoreticians that define them as subordinates charged with the task of "applying" a theory or that capture their practice as an illustration of a theory. [emphases and parentheses added]

The crucial point here for the obligations and responsibilities for pedagogical practice, to repeat, is the focus upon the notion of cosmos viewed as the 'unknown constituted by divergent worlds and the articulations of which they may be possible'. In other words, the focus upon the potentialities of practices and what in Scene 1 I referred to as the instauration of cosmicities. For Stengers, 'approaching a practice means approaching it as it diverges' (2005b, p. 184) and by employing the conceptual tool of an ecology of practices there is no intention to, 'describe practices "as they are" but as they may become' (2005b, p. 186). Given some leeway, we might begin to take on board the notion of cosmopedagogies that are called for in those 'concrete situations where practitioners, such as teachers, operate' and who are able to respond to the task of responding to how something matters for a particular learner and the possibilities that may ensue.

Cosmopedagogies therefore constitute pedagogical adventures that require an obligation not to represent practice 'according to' but to invent ways of responding through 'learning with' and 'composing with'. This is not to reject the structures and formatting of institutional education but to slow down, when approaching the diverse articulations of practice and what they might be capable of. We might say that cosmopedagogies of taking care are always emergent in their task of responding to the divergence of practices but as equally able to examine the limitations of established pedagogical frameworks when these are challenged. An ethics and politics of pedagogical work cannot work exclusively from established boundaries that anticipate compliance to them but have to negotiate strategies commensurate with the difference and divergence it faces.

The notion of trust plays a crucial role in cosmopedagogies. Trust by a teacher who is willing to acknowledge that he or she does not know what a student is capable of but is willing to explore and support, and trust by the student in their teacher believing in her capabilities even though these may remain inchoate. The pedagogical trust of a teacher when encountering diverse practices brings to mind Whitehead's point in *Process and Reality* (1979, p. 338) when he describes the responsibility of philosophy of not neglecting 'the multifariousness of the world' and avoiding the temptation to rush in and make judgements that may silence or devalue those whose practice does not conform.

Stengers (2005b) adopts this notion of difference as well as the need to slow down in her notion of the ecology of practices that places emphasis upon what is important, what is of value for a particular practice. This question of value is also reiterated by Whitehead (1938, p. 116) in his last book, *Modes of Thought,* where he writes, 'Have a care, here is something that matters.' The ecology of each practice, along with the more extensive notion of an ecology of practices, and the attainment of value or importance, are crucial leitmotifs for the deployment of cosmopedagogies in responding to each student's practice and its particular attainment of importance and subsequent potential. We might say that cosmopedagogies have to work with the pluralism of what is presented but also to engage in a speculative adventure in relation to that part of the student's process of learning that lies in the future. And, equally, to repeat, to interrogate the limitations of pedagogic schemas when these are challenged. In the words of Souriau, cosmopedagogies are always incomplete works-to-be-made.

Cosmopedagogies imply the scenario painted by the title of Martin Savransky's book, *Around the Day in Eighty Worlds* (2021). We might apply this pluralistic journey to consider the task of negotiating a journey around the classroom or studio as a journey alongside multifarious worlds. He quotes an Andean saying which 'teaches that in order to grasp the universe in its totality' we must 'begin to notice each of its components, one by one, until we fall asleep' (p. 25). This relates closely to the teacher's situated art of noticing the multifarious nature of pedagogical work and the worlds that may be possible. It reverses the approach to pedagogic work which is guided by overarching guidelines or competences that reduce the pluriverse of practices to their criteria. Though the following quotation by Susan Buck-Morss (2010, pp. 76–77) may seem a little oblique it does hit the target of institutional apparatuses that marginalise or even ignore those practices that 'do not fit' and the capabilities that these may become capable of.

> There is a blindness to institutionalised education that passes down the authority of tradition, a mental timidity, born of privilege or just plain laziness, that cloaks itself in the heavy bombast of cultural heritage and historic preservation. It generates enormous resistance to trespassing conceptual boundaries or exceeding the limits of present imagination, rewarding instead the virtues of scholastic diligence, disciplinary professionalism and elitist erudition, all escape routes from the pragmatic necessity of confronting the new. Indeed extreme discomfort is caused by the truly new, the truly 'contemporary,' that which Nietzsche called the 'untimely'—those aspects of the present moment that simply do not fit our established traditions or modes of understanding.

This blindness is often reinforced by what Deleuze (1986, p. 106) refers to as *la bêtise* or stupidity. He writes, in *Nietzsche and Philosophy:*

> When someone asks 'what's the use of philosophy?' the reply must be aggressive, since the question tries to be ironic and caustic. [...] It [philosophy] is useful for harming stupidity [*la bêtise*], for turning stupidity into something shameful. (p. 106)

We might take the notions of stupidity and shame and apply them to institutional modes of practice in education that turn a blind eye to supporting divergence and difference whilst declaring the opposite through maxims such as, 'every child matters' or 'no child left behind'. We witness their stupidity and shame in the dire state of educational provision for special needs, the failure to provide adequate resourcing and support for neurodiversity. Indeed, such failure stymies the opportunity to widen pedagogical practices. Stupidity for Deleuze, and for Stengers who resists its impact more globally and politically, does not refer to individual psychological states but rather to something that grips or captures an epoch, that seizes hold of people. For Deleuze the task of philosophy as stated above is to harm such stupidity and turn it into something shameful. Has our education service, driven by economic imperatives, established traditions, an obsession with standards and accountability been captured by what both Stengers and Deleuze refer to as stupidity? In her book, *In Catastrophic Times* (2015, p. 117), Stengers writes:

> What I will name stupidity cannot be reduced to a type of psychological weakness. It will not be said that "people are stupid" as if it was a matter of some personal defect. Stupidity is something about which it will be said instead *that it seizes hold of certain people*. And in particular it seizes hold of those who feel themselves in a position of responsibility and who then become what I call our guardians.

It is the stupidity and shame attached to the blind-eye that is turned towards 'other' practices and their potential, to the 'untimely', in order to continue in the arrogant confidence of its current groove, rejecting any alternatives, so ready and confident to speak on behalf of these 'others', that must be resisted. This problem was also echoed by Whitehead (1967, p. 197) in his criticism of the method of training of professionals in the nineteenth century, developing what he called 'minds in a groove' that treated those factors or practices outside the groove as irrelevant or unimportant. When we consider our current international epoch of educational policy and directives, driven by standards, competences, assessment etc., it seems incapable of evaluating the situated importance of each practice, that which gives a sense of importance to a particular practice, and is likely to disqualify those that 'do not fit' (Fig. 20.1).

The pedagogical challenge for cosmopedagogies that are required to respond to the divergence and difference of a student's learning practices is to be able to create modes of practice that enable it to deal with such diversity. This may involve a teacher having to decolonise his or her

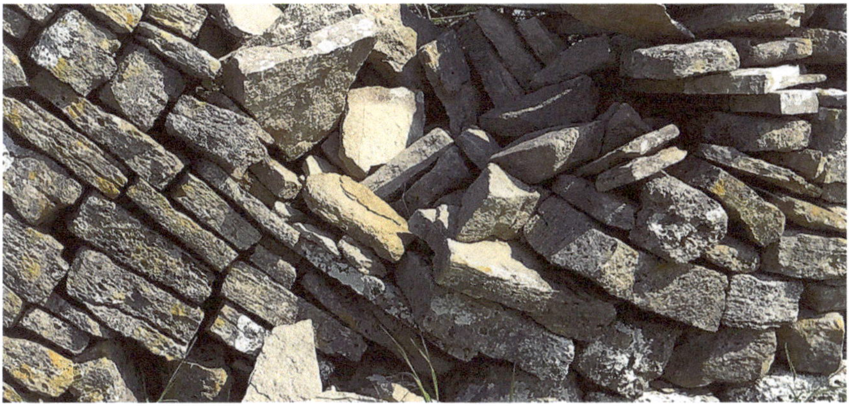

Fig. 20.1 Material collectivity

pedagogical frameworks in order to enable those differences to flourish whilst at the same time expanding the teacher's understanding of practice in its specific domain. Such processes of pedagogical transformation as they arise in concrete situations of practice seems to require, as Savransky (2021, p. 43) puts it, 'hesitating about our own metaphysical foundations [and] pluralizing our own concept of reality', in encounters with those practices that 'do not fit' expectations. It is in such moments of hesitation that we may become aware of cosmos and its as yet unknown potentials. The metaphysical determination of educational policy, 'this is how the world is and should become' can be contrasted with a metaphysical indetermination of cosmicities in-the-making and what they may be capable of, which is the concrete reality of pedagogic work that is obligated to respond to the pluralism of learning pathways without imposing a specific roadmap, but which remains open to the ontogenesis of such pathways. Such pedagogies are therefore always being made and re-made. It is a pedagogy which, in the words of William James (1996, p. 73), 'keeps the doors and windows open'. It is a cosmopedagogy that attempts to work with the complex of an ecology of learning practices. This is not an easy task considering the pressures and demands of institutional contexts of education today.

Competing Interests (4) The author has no conflicts of interest to declare that are relevant to the content of this chapter.

References

Buck-Morss, S. (2010). The second time as farce...historical pragmatics and the untimely present. In C. Douzinas & S. Zizek (Eds.), *The idea of communism* (pp. 67–80). Verso.

de Castro, V. (1998). Cosmological deixis and Amerindian perspectivism. *The Journal of the Royal Anthropological Institute, 4*(3), 469–488.

Deleuze, G. (1986). *Nietzsche and philosophy*. Continuum.

Deleuze, G., & Guattari, F. (1994). *What is philosophy?* Verso.

James, W. (1996). *Some problems of philosophy*. Nebraska University Press.

Lapoujade, D. (2021). *The lesser existences: Etienne Souriau, an aesthetics for the virtual*. University of Minnesota Press.

Penketh, C. (2023). *A history of disability and art education*. Routledge.

Roberts, T., Lapworth, A., & Dewsbury, J. D. (2022). From 'world' to 'earth': non-phenomenological subjectivity in Deleuze and Guattari's geophilosophy. *Subjectivity, 15*, 135–151.

Savransky, M. (2021). *Around the day in eighty worlds*. Duke University Press.

Souriau, E. (1938). *Avoir une ame: Essai sur les existences virtuelles*. Societe d'Edition Les Belles Lettres.

Souriau, E. (2015a). *The different modes of existence*. Univocal Publishing.

Souriau, E. (2015b). Of the mode of existence of the work to be made. Appendix to *The different modes of existence* (pp. 219–240). Universal Publishing

Stengers, I. (2005a). The cosmopolitical proposal. In B. Latour & P. Weibel (Eds.), *Making things public* (pp. 994–1003). MIT Press.

Stengers, I. (2005b). Introductory notes of an ecology of practices. *Cultural Studies Review, 11*(1), 183–196.

Stengers, I. (2006). *Whitehead and science: From philosophy of nature to speculative cosmology*. Lecture at McGill University. Retrieved May 31, 2024, from https://www.mcgill

Stengers, I. (2015). *In catastrophic times: Resisting the coming barbarism*. Open Humanities Press and Meson Press.

Whitehead, A. N. (1979). *Process and reality*. Free Press.

Whitehead, A. (1938). *Modes of thought*. Free Press.

Whitehead, A. N. (1967). *Science and the modern world*. Free Press.

Correction to: Ecologies in Practice and Learning

Miranda Matthews

Correction to:
M. Matthews (ed.), *Ecologies in Practice and Learning: Arts Interventions in the Earth Crisis*, Palgrave Studies in Educational Futures, https://doi.org/10.1007/978-3-031-82338-1

This book was inadvertently published with errors in various chapters. These errors have been rectified in this revised version.

The updated version of this book can be found at
https://doi.org/10.1007/978-3-031-82338-1

M. Matthews (ed.), *Ecologies in Practice and Learning: Arts Interventions in the Earth Crisis*, Palgrave Studies in Educational Futures, https://doi.org/10.1007/978-3-031-82338-1_21

Index

M. Matthews (ed.), *Ecologies in Practice and Learning: Arts Interventions in the Earth Crisis*, Palgrave Studies in Educational Futures, https://doi.org/10.1007/978-3-031-82338-1

T

FSC
www.fsc.org
MIX
Papier | Fördert
gute Waldnutzung
FSC® C083411

Kolibri 360 GmbH
Ferdinand-Jühlke-Straße 7
99095 Erfurt, Deutschland
produktsicherheit@kolibri360.de